Robert William Griffin

The Parabola, Ellipse and Hyperbola, Treated Geometrically

Robert William Griffin

The Parabola, Ellipse and Hyperbola, Treated Geometrically

ISBN/EAN: 9783744750677

Printed in Europe, USA, Canada, Australia, Japan

Cover: Foto ©berggeist007 / pixelio.de

More available books at **www.hansebooks.com**

DUBLIN UNIVERSITY PRESS SERIES.

THE PARABOLA, ELLIPSE, AND HYPERBOLA,

TREATED GEOMETRICALLY.

BY

ROBERT WILLIAM GRIFFIN, A.M., LL.D.,

EX-SCHOLAR, TRINITY COLLEGE, DUBLIN.

DUBLIN: HODGES, FOSTER, & FIGGIS, GRAFTON-ST.
LONDON: LONGMANS, GREEN, & CO., PATERNOSTER-ROW.

1879.

PREFACE.

Long experience has proved to me that it would be a great practical advantage for the general class of students to acquire a knowledge of even the elementary Geometrical properties of the Parabola, Ellipse, and Hyperbola—a knowledge which may be attained, in a very short time, by any one acquainted with the six Books of Euclid, though many have neither the ability nor the perseverance necessary for the Analytical investigation.

In the following Treatise I have endeavoured to demonstrate, on strictly geometrical principles, the most useful properties of these curves. In the definition of a tangent I have avoided the notion of a limit, adopting, in preference, Euclid's definition; and the demonstrations depending thereon will, I think, be found somewhat new and interesting.

The similar properties of the three curves will be found to be treated in such a manner as to need scarcely any change of either words or notation.

In Chap. III. Proposition XIX. *et seq.* and Corollaries, I have traced the analogy between the properties of conjugate diameters of the Hyperbola, and those of the Ellipse, further than has been done in any other Treatise.

To the Board of Trinity College I owe my grateful thanks, for extending to me that liberal support with which they have ever shown themselves ready to assist the humblest efforts.

ROBERT WM. GRIFFIN.

19, TRINITY COLLEGE,
September, 1879.

CONTENTS.

CHAPTER I.

	PAGE
The Parabola,	1
Problems on the Parabola,	45

CHAPTER II.

The Ellipse,	46
Problems on the Ellipse,	110

CHAPTER III.

The Hyperbola,	111

Appendix,	178

CHAPTER I.

THE PARABOLA.

DEFINITIONS.

A PARABOLA is the curve traced out by a point, which moves in such a way that its distance from a fixed point is always equal to its perpendicular distance from a fixed right line.

The fixed point is called the *Focus*, and the fixed right line the *Directrix*.

Any right line perpendicular to the Directrix is called a *Diameter*.

The right line drawn through the Focus perpendicular to the directrix is called the *Axis*, and the point at which it meets the curve the *Vertex*.

A right line which meets the curve, and, being produced, does not cut it, is called a *Tangent*.

A right line drawn through any point on the curve perpendicular to the tangent at that point is called a *Normal*.

If a right line be drawn parallel to any tangent, the part intercepted on it, between the curve and a diameter passing through the point of contact of the tangent, is called an *Ordinate* to that diameter.

The part intercepted on any diameter between the ordinate and the curve is called the *Abscissa*.

2   *The Parabola.*   [CHAP. I.

The right line joining any two points on the curve is called a *Chord*.

The chord drawn through the Focus parallel to the ordinates to any diameter is called the *Parameter* of that diameter.

The chord drawn through the focus at right angles to the axis is called the *Latus rectum*.

Proposition I.

The focus and directrix of a parabola being given, to determine any number of points on the curve.

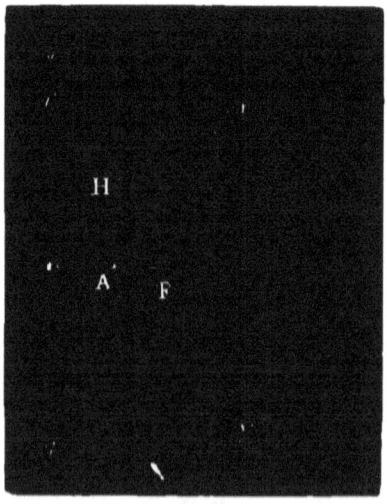

Fig. 1.

Let F be the focus, and Oy the directrix.
Draw $FO \perp$ the directrix; bisect FO at A.
Then OF is the axis, and A the vertex of the curve.

On the directrix take any point p; join Fp; draw $AH \perp FO$; through H draw $HP \perp Fp$, to meet pP drawn $\perp$ the directrix; join PF.

Then $\triangle^s\ FHP$ and pHP are equal;   (4 I. Euclid.)
$\therefore\ PF = Pp$. Hence P is a point on the curve.

CHAP. I.] *The Parabola.* 3

In like manner, by joining the focus with other points on the directrix, any number of points on the curve may be determined.

Cor. 1.—If Oq be taken $= Op$, another point Q may be found in a similar manner to P, which will be at the same distance both from the axis and the directrix; hence the curve is symmetrical with regard to the axis.

Cor. 2.—Since $Pp = Qq$;

$$\therefore FP = FQ, \text{ also } \angle OFP = \angle OFQ;$$

∴ right lines drawn from the focus to the curve, making equal angles with the axis, are equal.

Cor. 3.—The Latus rectum is equal to four times the distance of the focus from the vertex.

Fig. 2.

Draw PP' through the focus $\perp$ the axis; and $Pp \perp$ the directrix.

Then $\qquad PP' = 2PF \qquad\qquad (Cor.\ 2.)$

$\qquad\qquad\quad = 2Pp$

$\qquad\qquad\quad = 2FO = 4FA.$

Cor. 4.—If any parallel qR be drawn to the axis between the vertex A and any point P on the curve, the segment qQ intercepted on this line between the curve and the directrix is less than the segment qR intercepted on it between the line AP and the directrix; and, therefore, the curve is convex towards the directrix.

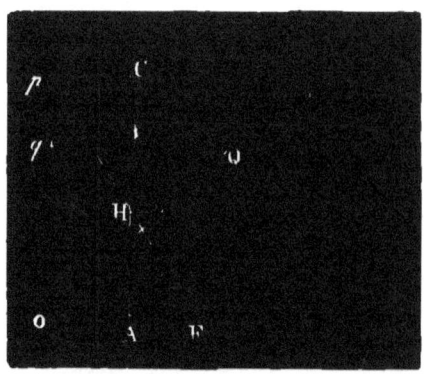

Fig. 3.

Since $\angle FHP$ is a right $\angle$; $\therefore \angle AHF + \angle CHP = 90°$.

Hence $\triangle AHF$ is similar to $\triangle CPH$;

$\therefore FA : AH = HC : CP$;

$\therefore FA \times CP = AH \times HC$.

But $FA = AO$; $\therefore FH = Hp$; $\therefore AH = HC$;

$\therefore FA \times CP = AH^2 = \tfrac{1}{4}AC^2$.

Similarly, $FA \times DQ = \tfrac{1}{4}AD^2$;

$\therefore CP : DQ = AC^2 : AD^2$

$ = CP^2 : DR^2$; (Similar $\triangle^s$.)

$\therefore CP^2 : CP \times DQ = CP^2 : DR^2$;

$\therefore CP \times DQ = RD^2$;

$\therefore DQ : RD = RD : CP = AD : AC$. (Similar $\triangle^s$.)

But by hyp., $AD < AC$; $\therefore DQ < DR$; $\therefore qQ < qR$.

CHAP. I.] *The Parabola.* 5

Cor. 5.—Also if a perpendicular QN be drawn to the axis between the vertex A and any point P on the curve, the segment QN intercepted on this line between the curve and the axis is greater than the segment SN intercepted on it between the line AP and the axis; and, therefore, the curve is concave towards the axis.

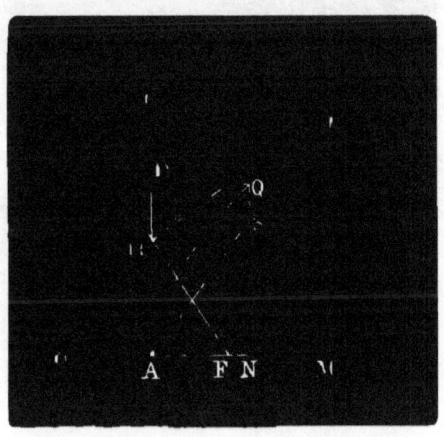

Fig. 4.

For $\qquad CP : DQ = AC^2 : AD^2 ;$ \qquad (*Cor.* 4.)

$\therefore AM : AN = PM^2 : QN^2 ;$

$\therefore PM : SN = PM^2 : QN^2 ;$ \qquad (Similar $\triangle^s$.)

$\therefore PM^2 : PM \times SN = PM^2 : QN^2 ;$

$\therefore PM \times SN = QN^2 ;$

$\therefore PM : QN = QN : SN ;$

but it was proved above that

$$PM^2 : QN^2 = AM : AN,$$

and by hyp.,

$AN < AM ; \therefore QN < PM,$ and $\therefore SN < QN.$

Proposition II.

The distance of any point from the focus is greater or less than its distance from the directrix according as the point is on the convex or concave side of the curve.

Fig. 5.

First let the point Q be on the convex side; join QF, and let the perpendicular Qp on the directrix be produced to meet the curve in P; join PF.

Then $\qquad QF + QP > PF;$ \qquad (20 I. Euclid.)

$\therefore \ QF + QP > Pp;$

$\therefore \qquad QF > Qp.$

Next let Q' be on the concave side; join $Q'F$.

Then $\qquad Q'F < Q'P + PF;$ \qquad (20 I. Euclid.)

$\therefore \ Q'F < Q'P + Pp;$

$\therefore \ Q'F < Q'p.$

Cor.—Conversely a point will be on the concave or convex side of a parabola according as its distance from the focus is less or greater than its perpendicular distance from the directrix.

Proposition III.

1°. The line which bisects the angle between that drawn from any point on a parabola to the focus, and that drawn perpendicular to the directrix, falls wholly without the curve.

2°. Any other line drawn through the point will cut the curve.

Fig. 6.

1°. Let PH bisect the angle between PF and Pp, the perpendicular on the directrix. Take any point Q on PH; join QF, and draw $Qq \perp$ the directrix.

Then the $\triangle^s$ QPF and QPp are equal; (4 I. Euclid.)

$$\therefore QF = Qp,$$

but $Qp > Qq;$ (19 I. Euclid.)

$$\therefore QF > Qq;$$

∴ Q is a point without the curve. (*Cor.* Prop. II.)

2°. Let PK be any line other than the bisector of the $\angle FPp$.

Fig. 7.

Draw PL, making the $\angle KPL = \angle KPF$; cut off $PL = PF$. Through L draw $QLq \perp$ the directrix; join QF.

Then the $\triangle^s FPQ$ and LPQ are equal; (4 I. Euclid.)

$$\therefore QL = QF;$$

$$\therefore QF < Qq;$$

∴ the point Q is on the concave side of the curve.
(*Cor.* Prop. II.)

Cor. 1.—Hence the line which bisects the angle between that drawn from any point on the curve to the focus, and that drawn perpendicular to the directrix, is a *Tangent* to the curve.

Cor. 2.—The tangent at the vertex is perpendicular to the axis, and ∴ ordinates to the axis are also perpendicular to it.

Cor. 3.—The normal to a parabola at any point bisects the external angle between the line drawn from that point to the focus, and the perpendicular let fall on the directrix.

[CHAP. I.] *The Parabola.* 9

Proposition IV.

The locus of the foot of the perpendicular let fall from the focus on any tangent to a parabola is the tangent at the vertex.

Fig. 8.

Let PT be any tangent, and FT the perpendicular let fall from the focus. Draw $Pp \perp$ the directrix; join pT.

Then the $\triangle^s FPT$ and pPT are equal; (4 I. Euclid.)

$\therefore \angle pTP = \angle FTP = 90°$;

$\therefore Tp$ and TF are in directum. (14 I. Euclid.)

Also $Tp = TF$; $\therefore TA$ is $\parallel Op$. (2 VI. Euclid).

Hence TA is a tangent at the vertex.

(*Cor.* 2, Prop. III.)

Cor. 1.—If the vertex of a right $\angle FTP$ move along a fixed right line, while one leg passes through a fixed point F, the other leg will always touch a parabola, of which the fixed point is the focus, and the fixed right line the tangent at the vertex.

Cor. 2.—If a perpendicular Pp be let fall on the directrix from any point P on a parabola, the line Fp joining the foot of this perpendicular with the focus is $\perp$ the tangent at the point P.

Cor. 3.—By the aid of this Prop. we can draw a pair of tangents to a parabola from any point Q without it.

Fig. 9.

On the line joining Q with the focus as diameter describe a circle; the line joining the given point Q with the points of intersection of this circle with the tangent at the vertex will be the required tangents.

Note.—The circle described on QF as diameter will always intersect the tangent at the vertex.

For, bisect QF in O' and draw $O'N \perp AM$.

Then $Qq = QM + Mq = QM + AO = QM + AF = 2O'N$,

but $\quad QF > Qq$ (Prop. II.); $\therefore QF > 2O'N$ or $O'F > O'N$.

That is, the radius of the circle is always greater than the perpendicular let fall from its centre on the tangent at the vertex.

Proposition V.

Any line drawn through the focus is cut harmonically by the curve, the focus, and the directrix.

Fig. 10.

For $\quad PD : DQ = Pp : Qq \quad$ (Similar △^{s}.)

$\quad\quad\quad\quad\quad = PF : FQ$.

Hence DP, DF, and DQ are in harmonic proportion.

Cor. 1. $\quad Pp : FO : Qq = DP : DF : DQ$.

$\quad\quad\quad\quad\quad\quad\quad\quad\quad\quad\quad$ (Similar △^{s}.)

Hence Pp, FO, and Qq, are in harmonic proportion.

∴ $\dfrac{1}{Pp}$, $\dfrac{1}{FO}$, and $\dfrac{1}{Qq}$, are in Arith. Progression.

∴ $\dfrac{1}{Pp} + \dfrac{1}{Qq} = \dfrac{2}{FO}$; and ∴ constant.

Cor. 2.—Hence, also, $\dfrac{1}{FP} + \dfrac{1}{FQ}$ is constant.

Proposition VI.

The locus of the intersection of tangents to a parabola which cut at right angles is the directrix.

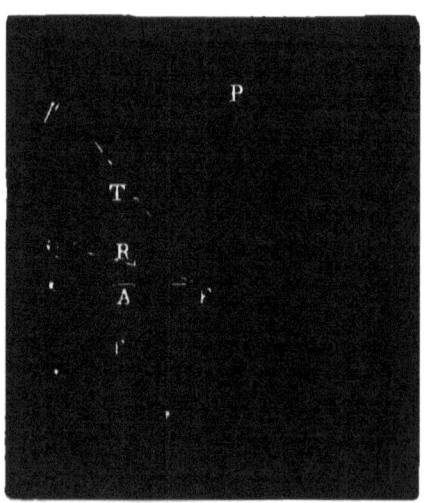

Fig. 11.

Let QP, QP', be tangents which cut at right angles. From the focus draw the perpendiculars FT, FT', on the tangents.

Then T and T' are points on the tangents at the vertex.
(Prop. IV.)

∴ TT' is ⊥ the axis. (*Cor.* 2, Prop. III.)

Also since $FTQT'$ is a rectangle, its diagonal QF is bisected at R.

Draw $\qquad QO \parallel TT'$.

Then $\qquad FA = AO$; (2 VI. Euclid.)

and $\quad QO$ being $\parallel TT'$, is ∴ ⊥ the axis.

Hence $\qquad$ ∴ QO is the directrix.

CHAP. I.]   *The Parabola.*   13

PROPOSITION VII.

If PN be a normal to a parabola at any point P, and PM the ordinate to the axis, the subnormal MN is of a constant length and equal to half the latus rectum.

Fig. 12.

Draw $Pp \perp$ the directrix.

Then Fp will be $\perp$ the tangent at P; and $\therefore \parallel PN$.
(*Cor.* 2, Prop. IV.)

Hence $\triangle FpO = \triangle NPM$;   (26 I. Euclid.)

$\therefore MN = FO = 2FA = \tfrac{1}{2}$ Latus rectum.
(*Cor.* 3, Prop. I.)

Def.—The portion of the axis intercepted between the normal at any point on the curve, and the ordinate to the axis drawn through the same point, is called the *Subnormal*.

The portion of the axis intercepted between the tangent at any point on the curve, and the ordinate to the axis drawn through the same point, is called the *Subtangent*.

Proposition VIII.

If a tangent at any point P on a parabola meet the axis produced in T, and the ordinate PM be drawn, then the subtangent MT will be bisected at the vertex.

Fig. 13.

Draw $Pp \perp$ the directrix; join FP.

Then $\angle FPT = \angle TPp = \angle PTF$;

$\therefore FT = FP = Pp = MO$;

but $AF = AO$;

$\therefore AT = AM$.

Cor. 1.—If PN be a normal, then $FT = FN$.

For $FT = MO = MF + FO = MF + MN = FN$.

(Prop. vii.)

Cor. 2. $\quad \angle PFN = 2 \angle PTF$.

Proposition IX.

The rectangle under the Latus rectum and the abscissa is equal to the square of the ordinate.

Fig. 14.

Draw PT the tangent and PN the normal at the point P.

Then $PM^2 = TM \times MN$; (8 VI. Euclid.)

but $TM = 2AM$; and $MN = 2AF$;

(Props. VII. and VIII.)

$\therefore PM^2 = 2AM \times 2AF = 4AF \times AM$.

Cor. 1.—The squares of the ordinates to the axis are proportional to the abscissas.

Cor. 2.—The square of the tangent $= 4FP \times AM$.

For $TP^2 = TN \times TM$ (8 VI. Euclid.)

$= 2TF \times 2AM$

$= 4FP \times AM$.

Cor. 3.—The square of the normal $= 4AF \times FN$.

For $PN^2 = TN \times NM = 2FN \times 2AF = 4AF \times FN$.

16 The Parabola. [CHAP. I.

PROPOSITION X.

The locus of the intersection of any tangent to a parabola with the line drawn through the focus perpendicular to the radius vector drawn from the focus to the point of contact of the tangent is the directrix.

Fig. 15.

Let PT be the tangent and FT the $\perp$ the radius vector FP.

Draw $TO \perp$ the axis, and $Pp \perp TO$ produced.

Then $\quad\quad\quad \triangle TFP = \triangle TpP;\quad\quad$ (26 I. Euclid.)

$\quad\quad\therefore Pp = PF;$

hence OT is the directrix.

Cor. 1.—Conversely, if from any point on the directrix a tangent be drawn to a parabola, the line joining that point to the focus is perpendicular to the radius vector drawn from the focus to the point of contact of the tangent.

Cor. 2.—Tangents at the extremities of a focal chord intersect at right angles on the directrix.

For $\quad\angle FTP = \angle pTP$ and $\angle FTQ = \angle qTQ$;

$\quad\therefore QTP = \tfrac{1}{2}(\angle FTp + \angle FTq)$

$\quad\quad\quad = $ a right $\angle$.

Proposition XI.

If any chord PP' of a parabola cut the directrix in D, then FD is the external bisector of the angle PFP'.

Fig. 16

Draw Pp, $P'p'$ ⊥ the directrix.

Then $\qquad P'F : PF = P'p' : Pp$

$\qquad\qquad\qquad = P'D : DP;\qquad$ (Similar $\triangle^s$.)

∴ FD is the bisector of the ∠ PFQ.

(3 VI. Euclid.)

Cor. 1.—Hence, being given the focus F and two points P, P', on a parabola, we can find the directrix and axis.

For, draw FD bisecting the ∠ between FP and FP' produced; the point D where this line meets PP' produced is a point on the directrix. The tangent drawn from D to the circle described with the centre P and the radius PF, will be the directrix, and the perpendicular let fall on it from the focus, the axis.

If a tangent to a curve be defined as *the line joining two indefinitely near points on the curve*, it will follow immediately from this Prop. that the right line drawn from the focus to the point of intersection of any tangent with the

18   *The Parabola.*   [CHAP. I.

directrix is perpendicular to the radius vector drawn from the focus to the point of contact of the tangent; also, that the tangent makes equal angles with the lines drawn from the point of contact to the focus and the perpendicular to the directrix.

For when P is indefinitely near to P', PD becomes a tangent at the point P, and the $\angle PFP'$ is indefinitely small, and $\therefore PFQ = 180°$; but $\angle PFD = \tfrac{1}{2} \angle PFQ$, $\therefore \angle PFD = 90°$.

Also, $\qquad \triangle PFD = \triangle PpD;$
$\qquad \therefore \angle DPF = \angle DPp.$

Proposition XII.

If two fixed points PP' on a parabola be joined with a third variable point O on the parabola, the segment pp' intercepted on the directrix by the produced chords subtends a constant angle at the focus.

Fig. 17.

Since (Prop. XI.) Fp is the bisector of the $\angle$ between FO and FP produced;

$$\therefore \angle OFp + \tfrac{1}{2} \angle OFP = 90°;$$

also, since (Prop. XI.) Fp' is the bisector of ∠ between FO and FP' produced;

$$\therefore \angle OFp' + \tfrac{1}{2}\angle OFP' = 90°;$$
$$\therefore \angle OFp + \tfrac{1}{2}\angle OFP = \angle OFp' + \tfrac{1}{2}\angle OFP';$$
$$\therefore \angle OFp - \angle OFp' = \tfrac{1}{2}\angle OFP' - \tfrac{1}{2}\angle OFP;$$
$$\therefore \angle pFp' = \tfrac{1}{2}\angle PFP' \text{ and } \therefore \text{ constant.}$$

Cor.—Hence the anharmonic ratio of the pencil formed by joining four fixed points on a parabola to any fifth variable point is constant.

For $O \cdot PP'P''P''' = O \cdot pp'p''p''' = F \cdot pp'p''p'''$.

Proposition XIII.

The line joining the focus to the point of intersection of any two tangents to a parabola bisects the angle between the radii vectores drawn from the focus to the points of contact of the tangents.

Fig. 18.

Let QP, QP' be tangents. Join PF, $P'F$, QF. Draw $P'p'$, $Pp \perp$ the directrix. Join Qp, Qp'.

Then $\triangle^s QFP'$ and $Qp'P'$ are equal; (4 I. Euclid.)
$\therefore QF = Qp'$, also $\angle QFP' = \angle Qp'P'$.
Again, $\triangle^s QFP$ and QpP are equal; (4 I. Euclid.)
$\therefore QF = Qp$, and $\angle QFP = \angle QvP$.
Hence, $Qp = Qp'$, $\therefore \angle Qpp' = \angle Qp'p$, $\therefore \angle QpP = \angle QpP'$;
$\therefore \angle QFP = \angle QFP'$.

Cor. 1.—If Qq be drawn $\parallel$ the axis and produced it will bisect the chord PP'.

For $\triangle^s Qpq$ and $Qp'q$ are equal. (26 I. Euclid.)
$\therefore pq = qp'$; but Pp, Qq, and $P'p'$ are parallel;
$\therefore PM = MP'$.

Cor. 2.—Since $QF = Qp = Qp'$; $\therefore$ the circle described will centre Q, and radius QF will pass through p and p'.

Cor. 3.—By the aid of *Cor.* 2 we can draw a pair of tangents to a parabola from a point Q without it.

For, with centre Q, and radius QF, describe a circle from the points where this circle cuts the directrix; draw perpendiculars meeting the curve in P and P'; the lines joining Q with P and P' will be the required tangents.

Cor. 4.— $\triangle^s P'QF$ and QPF are similar, and QF is a mean proportional between FP and FP'.

For $\angle FQP' = \angle P'Qp'$, $\angle FQP = \angle PQp$,
and $\angle pQq = \angle p'Qq$;
$\therefore \angle FQP' + PQq = 180°$;
but $\angle pPQ$ or $\angle FPQ + \angle PQq = 180°$; (29 I. Euclid.)
$\therefore \angle FQP' = \angle FPQ$,
also $\angle QFP = \angle QFP'$.
Hence $\triangle P'QF$ is similar to $\triangle QPF$;
$\therefore P'F : FQ :: FQ : FP$.

[CHAP. I.] *The Parabola.* 21

Proposition XIV.

The chord of contact of any two tangents to a parabola is an ordinate to the diameter passing through their point of intersection.

Fig. 19.

Let QP, QP', be two tangents, QM a parallel to the axis passing through their intersection, meeting the curve in V. At V draw a tangent meeting QP in R, and QP' in R'; join VP, VP'; through R and R' draw parallels to the axis.

Then VP will be bisected in S, and VP' in S'.
(*Cor.* 1, Prop. XIII.)

Hence QP will be bisected at R, and QP' at R';
(2 VI. Euclid.)

∴ RR' is $\parallel PP'$. (2 VI. Euclid.)

Hence PP' is an ordinate to diameter QM. (*Def.*)

Cor. $QV = VM$.

Proposition XV.

The angle at the focus, subtended by the segment intercepted on a variable tangent to a parabola by two fixed tangents, is equal to half the angle which the chord of contact of the fixed tangents subtend at the focus, and therefore constant.

Fig. 20.

Let QP, QP', be the fixed tangents, RS the variable; join F with P', S, T, R, and P.

Then $\angle SFT = \frac{1}{2} \angle TFP'$; also $\angle TFR = \frac{1}{2} \angle TFP$;
(Prop. xiii.)

∴ $\angle RFS = \frac{1}{2} \angle PFP'$, which is constant.

Cor.—The anharmonic ratio of the four points in which a variable tangent to a parabola is cut by four fixed tangents is constant.

For the segments intercepted on the variable tangents subtend constant angles at the focus.

Proposition XVI.

The circle circumscribing the triangle formed by any three tangents to a parabola passes through the focus.

CHAP. I.] *The Parabola.* 23

For $\angle PFM = 2 \angle PBF$,   (*Cor.* 2, Prop. VIII.)
and $\angle P'FM = 2 \angle P'DF$;   (*Cor.* 2, Prop. VIII.)
$\therefore \angle P'FM - \angle PFM = 2 \angle P'DF - 2 \angle PBF$,
or $\angle P'FP = 2 \angle BQD$;
$\therefore \angle SFR = \angle BQD$.   (Prop. XV.)

Hence a circle will pass through $SQRF$.

PROPOSITION XVII.

If a tangent at any point P on a parabola meet the axis in T and P, be joined with the vertex, the area of the triangle PAT thus formed will be equal to the area of the triangle PAp, formed by drawing Pp parallel to the axis, to meet the tangent at the vertex.

Fig. 21.

Draw the ordinate PM.
Then, since $TA = AM$;   (Prop. VIII.)
$\therefore \triangle TAP = \triangle MAP = \triangle PAp$.
(38 I. Euclid.)
Cor. $\triangle TMP$ = parallelogram $PMAp$.

Proposition XVIII.

If QG be an ordinate drawn to any diameter Pp, from any point Q on a parabola cutting the axis in L, and QS be an ordinate drawn to the axis from the same point, AH a tangent at the vertex, then the triangle QLS = parallelogram $KSAH$.

Fig. 22.

Draw PT the tangent at P.

Then $\triangle$s QLS and PTM are similar;

$\therefore \triangle QLS : \triangle PTM = QS^2 : PM^2$ (19 VI. Euclid.)

$\qquad = AS : AM$ (Cor. 1, Prop. IX.)

$\qquad = \square\, ASKH : \square\, AMPH$;

but $\triangle PTM = \square\, AMPH$; (Cor. Prop. XVII.)

$\therefore \triangle QLS = \square\, KSAH$.

Cor. 1.—In like manner it may be proved that

$$\triangle Q'S'L = \square\, AS'K'H.$$

Cor. 2.—The chord QQ' is bisected by the diameter Pp.

[CHAP. I.] *The Parabola.* 25

Since $\triangle QLS = \square KSAH$; $\therefore \triangle QGK =$ trap. $HALG$;
also $\triangle Q'LS' = \square K'S'AH$; $\therefore \triangle Q'GK' =$ trap. $HALG$;
$\therefore \triangle QGK = \triangle Q'GK'$; they are also similar;
$\therefore QG = Q'G$.

Cor. 3. $\triangle QGK = \square GLTP$.

For $\triangle TPM = \square AMPH$; (*Cor.* Prop. XVII.)
$\therefore \square TPGL =$ trap. $HALG = \triangle QGK$. (*Cor.* 2.)

Cor. 4.—Chords drawn parallel to the tangent at any point will be bisected by the diameter passing through that point.

This follows immediately from *Cor.* 2.

Proposition XIX.

If a line be drawn parallel to the chord of contact of two tangents, the parts intercepted on it between the curve and the tangents are equal.

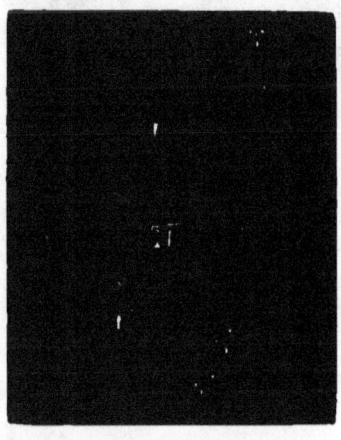

Fig. 23.

Draw the diameter through Q.
Then PP' is an ordinate to QM; (Prop. XIV.)

∴ SS', which is $\parallel PP'$, is also an ordinate;

∴ PP and SS' are bisected by QMN;
(*Cor.* 4, Prop. xviii.)
∴ $NT = NT'$.

Hence $\qquad TS = T'S'$.

Proposition XX.

The squares of the ordinates to any diameter are proportional to the abscissas.

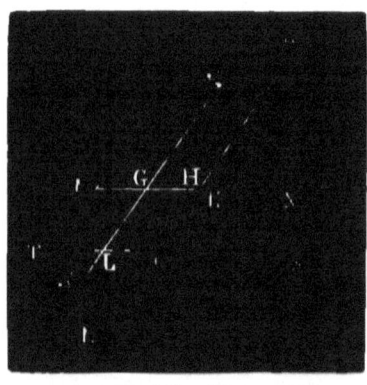

Fig. 24.

Let TS be the axis, PN any diameter, QG, RH ordinates $\parallel$ the tangent at P.

Then $\qquad \triangle QGK = \square PGLT$,
(*Cor.* 3, Prop. xviii.)

and $\qquad \triangle RHN = \square PHCT$;

∴ $\triangle QGK : \triangle RHN = \square PGLT : \square PHCT$.

Hence $\qquad QG^2 : RH^2 = PG : PH$.
(19 & 1 VI. Euclid.)

Proposition XXI.

The parameter of any diameter is four times the line joining the focus with the point where that diameter meets the curve.

Fig. 25.

Let QH be the focal ordinate to the diameter PH. Draw $AG \parallel QH$.

Then $QH^2 : AG^2 = PH : PG = TF : TA$;
$$= 4TF \times TF : 4TF \times TA;$$

but $AG^2 = PT^2 = 4TF \times TA$;

(Cor. 2, Prop. ix.)

$$\therefore QH^2 = 4TF \times TF;$$
$$\therefore QH = 2TF$$
$$= 2FP. \qquad \text{(Prop. viii.)}$$

Hence, parameter $= 4FP.$ (Cor. 4, Prop. xviii.)

Cor.—The abscissa intercepted on any diameter by a double focal ordinate $= \frac{1}{4}$ of the parameter of that diameter.

For $PH = TF = FP = \frac{1}{4}$ parameter.

Proposition XXII.

The square of the ordinate to any diameter is equal to the rectangle under the parameter of that diameter and the abscissa.

Fig. 26.

Let RS be any ordinate to the diameter PS, and QH the focal ordinate.

Then $\quad RS^2 : QH^2 = PS : PH;\quad$ (Prop. xx.)

but $\quad\quad\quad QH = 2FP \quad\quad$ (Prop. xxi.)

$\quad\quad\quad\quad\quad = 2FT \quad\quad$ (Prop. viii.)

$\quad\quad\quad\quad\quad = 2PH;$

$\therefore RS^2 : 4PH^2 = PS : PH$

$\quad\quad\quad\quad\quad = 4PH \times PS : 4PH \times PH;$

$\therefore RS^2 = 4PH \times PS$

$\quad\quad\quad = 4FT \times PS$

$\quad\quad\quad = 4FP \times PS$

$\quad\quad\quad = $ Parameter $\times$ Abscissa.

(Prop. xxi.)

Proposition XXIII.

If from any point on a parabola a perpendicular be let fall on any diameter, the square of this perpendicular will be equal to the rectangle under the Latus rectum and the abscissa.

Fig. 27.

Draw AC the tangent at the vertex; join CF.

Then $\triangle^s\ QHG$ and CAT are similar;

$\therefore QH^2 : QG^2 = CA^2 : CT^2$;

$\therefore QH^2 : 4FP \times PG = FA \times AT : FT \times TA$

(Props. XXII. and IV.)

$\qquad = FA : FT$ or FP

$\qquad = 4FA \times PG : 4FP \times PG$;

$\therefore QH^2 = 4FA \times PG =$ Latus rectum × Abscissa.

Note.—The parameter of any diameter is proportional to the square of the cosecant of the angle which its ordinates make with the axis,

$$\operatorname{cosec}^2 \angle QGH = \frac{QG^2}{QH^2} = \frac{4FP \times PG}{4FA \times PG} = \frac{4FP}{4FA};$$

$\therefore$ Latus rectum × $\operatorname{cosec}^2 \angle QGH = 4FP =$ parameter of diameter PH.

Proposition XXIV.

If a tangent be drawn at any point P on a parabola, and also an ordinate PM from same point to any diameter $Q'M$, the portion intercepted on that diameter between the tangent and the ordinated will be bisected by the curve.

Fig. 28.

Let QQ' be the chord ∥ the tangent at P.

Draw QH ⊥ the diameter passing through P,

and PS ⊥ $Q'M$ produced.

Then $GQ = GQ'$;   (Cor. 4, Prop. XVIII.)

∴ $QH = Q'H' = PS$;

but $QH^2 = $ Latus rectum $\times PG$,   (Prop. XXIII.

and $PS^2 = $ Latus rectum $\times Q'M$;

∴ $Q'M = PG = TQ'$.

Cor.—The tangents at the extremities of a double ordinate intersect on the diameter.

CHAP. I.] *The Parabola.* 31

Proposition XXV.

If any diameter intersect two parallel chords, the rectangles under the segments of these chords are proportional to the segments of the diameter intercepted between the chords and the curve.

Fig. 29.   Fig. 30.

Draw the diameter SM, bisecting the parallel chords PP', QQ'.

Then $PM^2 = 4SF \times SM,$ (Prop. XXII.)

and $VD^2 = 4SF \times SD$;

$\therefore PM^2 - VD^2 = 4SF \times DM,$

or $PB \times BP' = 4SF \times VB.$

Similarly $QB' \times B'Q' = 4SF \times VB'$;

$\therefore PB \times BP' : QB' \times B'Q' = VB : VB'.$

Cor.—If a chord be drawn through any point, the rectangle under the segments is equal to the rectangle under the parameter of the diameter bisecting the chord, and the part intercepted on a diameter passing through the point between the curve and the point.

Proposition XXVI.

If two chords of a parabola intersect, the rectangles under the segments are proportional to the parameters of the diameters bisecting the chords.

Fig. 31.   Fig. 32.

Let CM, DN, be the diameters bisecting the chords QQ', PP'. Draw the diameter BV through the intersection of the chords.

Then $PB \times BP' = 4FD \times VB$, (*Cor.*, Prop. xxv.)

and $QB \times BQ' = 4FC \times VB$; (*Cor.*, Prop. xxv.)

∴ $PB \times BP' : QB \times BQ' :: 4FD : 4FC$.

Cor. 1.—The rectangles under the segments of any two intersecting chords are proportional to the lengths of the parallel focal chords.

Cor. 2.—If two intersecting chords be parallel to two others, the rectangles under the segments of the one pair are proportional to the rectangles under the segments of the other pair.

[CHAP. I.] *The Parabola.* 33

Cor. 3.—If the points P and P' coincide, the line BP becomes a tangent. Hence, if from any point two tangents be drawn to a parabola, and from any other point two parallel chords, the rectangles under the segments of the chords are proportional to the squares of the tangents.

Proposition XXVII.

If TP and TQ be two tangents to a parabola, OC a line drawn parallel to either, cutting the other in O, the curve in A and C, and the chord of contact of the tangents in B, then OA, OB, and OC are in geometric proportion.

Fig. 33.

For
$$OA \times OC : OQ^2 = TP^2 : TQ^2$$
$$= OB^2 : OQ^2;$$

(Cor. 3, Prop. XXVI.)

$$\therefore OA \times OC = OB^2;$$
$$\therefore OA : OB :: OB : OC.$$

Proposition XXVIII.

If a circle intersect a parabola in four points, the common chords will be equally inclined to the axis.

Fig. 34.

Let PP', QQ' be the common chords intersecting in O.

Then $OP \times OP' : OQ \times OQ'$ = the ratio of the parallel focal chords, (*Cor.* 1, Prop. xxvi.)

but $OP \times OP' = OQ \times OQ'$;

(35 III. Euclid.)

∴ the focal chords drawn parallel to QQ' and PP' are equal.

Hence the focal chords, and ∴ the chords QQ' and PP' are equally inclined to the axis. (*Cor.* 2, Prop. i.)

Cor. 1.—In like manner it can be shown that PQ and $P'Q'$, also PQ' and $P'Q$ are equally inclined to the axis.

[CHAP. I.] *The Parabola.* 35

Proposition XXIX.

Any line OB drawn through the intersection of two tangents to a parabola is cut harmonically by the curve, the point, and the chord of contact of the tangents.

Fig. 35.

Through B and B' draw TS and $T'S' \parallel PQ$, the chord of contact of the tangents.

Then $BO : OB' = BT : B'T' = BS : B'S'$;
(Similar $\triangle^s$.)

$\therefore BO^2 : OB'^2 = BT \times BS : B'T' : B'S'$,

$= BT \times TA : B'T' : T'A'$,
(Prop. xix.).

$= TP^2 : PT'^2$,

$= BO^2 : O'B'^2$;

$\therefore BO : OB' = BO' : O'B'$.

Cor.—If through any point O without a parabola a line be drawn cutting the curve in B and B'; and OO' be taken equal to the harmonic mean between OB and OB', the locus of O' is a straight line, namely, the chord of contact of the tangents drawn from O.

PROPOSITION XXX.

If PM, QN be any diameters, BE a line drawn parallel to the ordinates to either, meeting the curve in D and QP in C, then BC, BD, and BE are in geometric proportion.

Fig. 36.

Draw the ordinate QG.

Then $\quad BE^2 : BD^2 = GQ^2 : BD^2$,

$\qquad\qquad\qquad = PG : PB, \qquad$ (Prop. xx.)

$\qquad\qquad\qquad = PQ : PC, \qquad$ (2 VI. Euclid.)

$\qquad\qquad\qquad = BE : BC,$

$\qquad\qquad\qquad = BE^2 : BE \times BC;$

$\therefore BD^2 = BE \times BC.$

Proposition XXXI.

If through any point P on a parabola lines POO', $Q'PQ$, be drawn parallel to any two adjacent sides DA, DC, of an inscribed quadrilateral, meeting opposite sides in O, O', and Q, Q', then $PO \times PO' : PQ \times PQ'$ in a constant ratio.

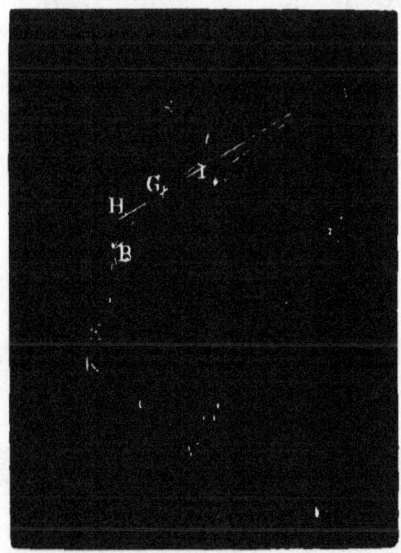

Fig. 37.

Through B and C draw BE and $CG \parallel$ to AD; join AG, and produce it to meet EB produced in H.

$$OL : BH = LA : AH$$
$$= O'D : DK. \quad \text{(Parallel lines.)}$$

Alternation $\quad OL : O'D = BH : DK.$

Also $\quad SC \text{ or } PO' : SQ = BK : KC. \quad$ (Similar $\triangle^s$.)

$\therefore OL \times PO' : O'D \times SQ = BH \times BK : DK \times KC;$

$\therefore OL \times PO' : PQ' \times SQ = EK \times BK : DK \times KC$*

$\qquad = O'V \times O'P : O'C \times O'D$

$\qquad = PL \times O'P : PS \times PQ'$;*

---

* The diameter which bisects the parallel chords GC and AD will bisect the chords BE and PV (Cor. 4, Prop. xviii.), also the parallel lines HK and LO'. Hence $BH = EK$, and $O'V = PL$.

38   The Parabola.   [CHAP. I.

$$\therefore OL \times PO' + PL \times O'P : PQ' \times SQ + PS \times PQ'$$
$$= EK \times BK : DK \times KC;$$
$$\therefore PO \times PO' : PQ \times PQ' = EK \times BK : DK \times KC.$$

Now it is evident that the points A, B, C, D being fixed, E is also fixed, and $\therefore EK \times BK : DK \times KC$ in a constant ratio, and $\therefore PO \times PO' : PQ \times PQ'$ in a constant ratio.

Proposition XXXII.

If from any point on a parabola lines PR, PR', PS, PS' be drawn to the sides of an inscribed quadrilateral, making with them any constant angles, then the rectangles under the lines drawn to the opposite sides will be in a constant ratio.

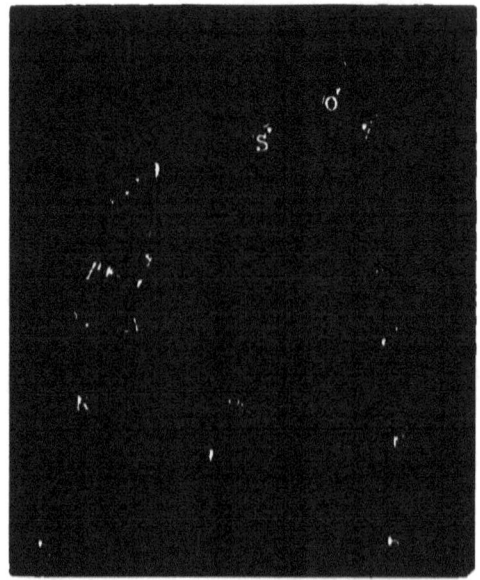

Fig. 38.

Take any other point p on the curve.

Through the points P, p, draw QPQ' and $pqq' \parallel DC$ and POO', $opo' \parallel AD$, also pr, pr', ps, ps', $\parallel PR$, PR', PS, PS', respectively.

Then $\qquad PR : pr = PQ : pq$, $\qquad$ (Similar $\triangle$'s).

and $\qquad PR' : pr' = PQ' : pq'$;

$\therefore PR \times PR' : pr \times pr' = PQ \times PQ' : pq \times pq'$.

Similarly it may be proved that

$\qquad PS \times PS' : ps \times ps' = PO \times PO' : po \times po'$;

but $\qquad PQ \times PQ' : PO \times PO' = pq \times pq' : po \times po'$;

$\therefore PR \times PR' : PS \times PS' = pr \times pr' : ps \times ps'$.

Cor. 1.—The rectangle under the perpendiculars let fall from any point of a parabola, on two opposite sides of an inscribed quadrilateral, is a constant ratio to the rectangle under the perpendiculars let fall on the other two sides.

Cor. 2.—If the points A and B coincide, also the points C and D, then the sides AB and CD become tangents, and the sides BC and AD coincide and become the chord of contact. Then the rectangle under the perpendiculars let fall from any point on a parabola on two fixed tangents is in a constant ratio to the square of the perpendicular let fall on their chord of contact.

Def.—Two curves are said to be *similar* and *similarly situated*, when a radius vector drawn from a fixed point in any direction to the first curve bears a constant ratio to the radius vector drawn from a fixed point in a parallel direction to the second curve.

Def.—Two curves are *similar*, when a radius vector drawn from a fixed point to the first curve bears a constant ratio to the radius vector drawn from a fixed point to the second curve, in a direction inclined at a constant angle to the former.

The two fixed points are called *centres of similarity*.

Proposition XXXIII.

All parabolas are similar figures.

Fig. 39.

Let F, f, be the foci, A, a, the vertices. Draw any two lines FP, fp, making equal $\angle^s$ with the axes. Draw tangents to the curves at P and p; let fall perpendiculars FT, ft, on the tangents from the foci; join TA, ta.

Then TA, ta, will be tangents at the vertices.

(Prop. IV.)

Hence also $\angle PFT = \frac{1}{2} \angle PFA$, and $\angle pft = \frac{1}{2} \angle pfa$;

(Prop. XIII.)

$$\therefore \angle PFT = \angle pft.$$

Hence $\triangle PFT$ is similar to pft.

Also $\triangle TFA$ is similar to $\triangle tfa$;

$\therefore PF : FT = pf : ft$, and $TF : FA = tf : fa$;

$\therefore PF : FA : : pf : fa$.

Alt. $PF : pf : : FA : fa$;

$\therefore$ The curves are similar.

Proposition XXXIV.

The area included between any ordinate PM, the abscissa AM, and the curve, is two-thirds of the parallelogram which has the ordinate and abscissa for its adjacent sides.

Fig. 40.

Draw tangents at A and P; join AP.

Since $TA = AM$, and $\therefore TH = HP$, (Prop. xxiv.)

$\therefore \triangle TAH = \frac{1}{2} \triangle TAP = \frac{1}{2} \triangle APM$.

Similarly, if through H we draw a diameter, and at the point Q where it meets the curve draw a tangent, we can prove that $\triangle HQh = \frac{1}{2} \triangle QGP$, and $\triangle HQh' = \frac{1}{2} \triangle QGA$. Continuing in this way to form new $\triangle^s$ by drawing diameters through h and h', we can prove that the areas of the exterior $\triangle^s$ formed by the tangents are the halves of the areas of the interior $\triangle^s$, formed by joining the points of contact with the extremities of the chords. The same will hold good though the number of $\triangle^s$ be increased indefinitely;

∴ the sum of all the exterior Δ^s = half the sum of the interior Δ^s.

Hence the area included between the curve and the lines AM and AP = $\frac{2}{3}$ area of $\Delta\ TPM$ = $\frac{2}{3}$ parallelogram $AMPB$.

Otherwise thus—

Fig. 41.

Draw the adjacent ordinate QN; join PQ and produce it; draw the tangent at A; draw $TC \parallel AB$, and $QD \parallel TM$.

Then, since PM is $\parallel AB$, the $\square\ QM = QC$.

Now when Q is indefinitely near P, PT becomes the tangent at P, and (Prop. xiv., *Cor*.) $AM = AT$.

$$\therefore\ \square\ BQ = \square\ BD;$$

$$\therefore\ \square\ QM = 2\ \square\ BQ.$$

If the parabolic arc AP be divided into an indefinite number of small parts, it can be similarly shown that each internal parallelogram is double the external, and hence the total internal area APM is double of the external area APB, and ∴ = $\frac{2}{3}$ of $\square\ AMPB$.

CHAP. I.]   *The Parabola.*   43

Def.—If a right-angled triangle be made to revolve round one of the sides containing the right angle, the surface generated by the hypothenuse is called a *right cone*, and the hypothenuse the *generating line*.

Proposition XXXV.

The *section* of a right cone by a plane parallel to the *generating line* is a parabola.

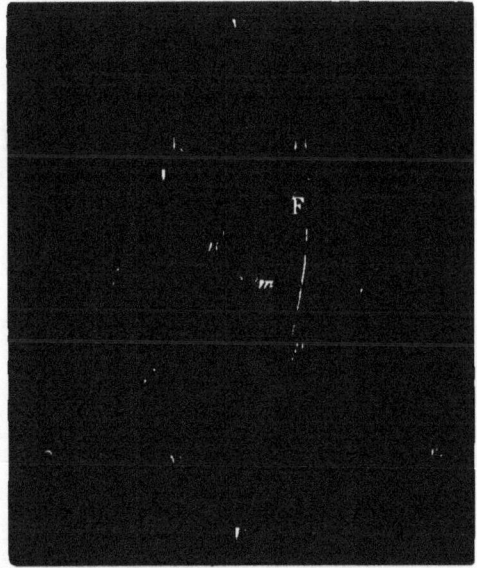

Fig. 42.

Let the plane BVE, drawn through the axis of the cone, perpendicular to the plane of the *section*, coincide with the plane of the paper, then both the *section* $P'AP$ and the base BPE will be perpendicular to the plane of the paper; therefore the line MP in which the *section* cuts the base is perpendicular to the plane of the paper, and, therefore, perpendicular to BE, the diameter of the base.

Hence $BM \times ME = MP^2$.   (35 III. Euclid.)

If now any other plane bpe be drawn parallel to the base, meeting the *section* in pm, it can similarly be shown that mp is perpendicular to be, and $\therefore bm \times me = mp^2$;

$$\therefore MP^2 : mp^2 = BM \times ME : bm \times me$$

$$= ME : me$$

$$= MA : mA. \qquad \text{(Similar } \Delta^s.)$$

Hence the section $P'AP$ is a parabola.

Cor. 1.—The latus rectum is a third proportional to VD and DA.

For $\quad VD : DA = AM : ME$

$$= AM \times MB : ME \times MB$$

$$= AM \times MB : MP^2$$

$$= AM \times DA : \text{Latus rectum} \times AM ;$$

(Prop. ix.)

$\therefore VD : DA = DA : \text{Latus rectum}.$

Cor. 2.—If from O a perpendicular be let fall on the section, the foot of this perpendicular will be the focus.

For $\quad \Delta^s\ VDO$ and OAF are equiangular;

$\therefore VD : 2DO = 2OA : 4AF;$

$\therefore VD : DA = DA : 4AF;$

$\therefore 4AF = 4$ Latus rectum. $\qquad$ (*Cor.* 1.)

Hence F is the focus. $\qquad$ (*Cor.* 2, Prop. i.)

Problems on the Parabola.

1. If from any point on a tangent to a parabola, a perpendicular be let fall on the radius vector drawn from the focus to the point of contact of the tangent, the segment intercepted on the radius vector, between the foot of the perpendicular and the focus, is equal to the perpendicular let fall from the point on the directrix.

2. The focus and a tangent being given, the locus of the vertex is a circle.

3. Find the locus of the point of intersection of any tangent to a parabola, with the line drawn from the focus, making a constant angle with the tangent.

If the vertex of a triangle of given species be fixed, while one base angle moves along a fixed right line, the locus of the other base angle will be a right line. (See also *Cor.* 1, Prop. IV.)

4. Given four tangents to a parabola, determine the focus and directrix. (See Prop. XVI.)

5. Normals at the extremities of a focal chord intersect on the diameter which bisects the chord. (See Prop. X.)

6. The circle described on any focal chord as diameter touches the directrix.

7. If a triangle ABC circumscribe a parabola whose focus is F, the lines drawn through A, B, C, perpendicular to FA, FB, and FC respectively, will pass through a point.

8. If a circle intersect a parabola in four points, the sums of the ordinates of the points of intersection on opposite sides of the axis are equal to each other.

9. Show how to cut from a cone a parabola having a given Latus rectum.

10. Prove that the lines joining the middle points of the sides of a triangle self-conjugate to a parabola are tangents to the curve. (See Props. XXIX. and XXIV.)

CHAPTER II.

The Ellipse.

DEFINITIONS.

An Ellipse is the curve traced out by a point, which moves in such a way that its distance from a fixed point is to its perpendicular distance from a fixed right line in a constant ratio $\epsilon : 1$ (ϵ being less than unity).

The fixed point is called the *Focus*, and the fixed right line the *Directrix*.

The right line drawn through the Focus perpendicular to the directrix is called the *Axis*.

The points at which the axis meets the curve are called the *Vertices*.

The middle point of the portion of the axis intercepted between the vertices is called the *Centre*.

Any line drawn through the centre is called a *Diameter*.

A right line which meets the curve, and being produced, does not cut it, is called a *Tangent*.

A right line drawn through any point on the curve perpendicular to the tangent at that point is called a *Normal*.

The right line joining any two points on the curve is called a *Chord*.

For Definitions of *Ordinate* and *Abscissa* see Definitions, Chapter I.

LEMMA.

If a right line AB be divided internally at O in any ratio, and externally at O' in the same ratio, and a circle be described on OO' as diameter, the right lines joining any point P on this circle with the extremities of the line AB will have the same ratio.

Fig. 1.

Bisect OO' in C; join CP, PO.

Then $AO' : O'B = AO : OB$;

∴ $AO' + AO : AO' - AO = O'B + OB : O'B - OB$;

∴ $2AC : 2OC = 2OC : 2BC$;

∴ $AC : CP = CP : CB$;

∴ $\triangle ACP$ is similar to $\triangle PCB$;   (6 VI. Euclid.)

∴ $\angle CPB = \angle CAP$;

but $\angle CPO = \angle COP$ (5 I. Euclid.)

     $= \angle OAP + \angle OPA$ (32 I. Euclid.)

     $= \angle CPB + \angle OPA$;

∴ $\angle BPO = \angle OPA$;

∴ $AP : PB = AO : OB$.   (3 VI. Euclid.)

Proposition I.

The focus, directrix, and eccentricity of an ellipse being given, to determine any number of points on the curve.

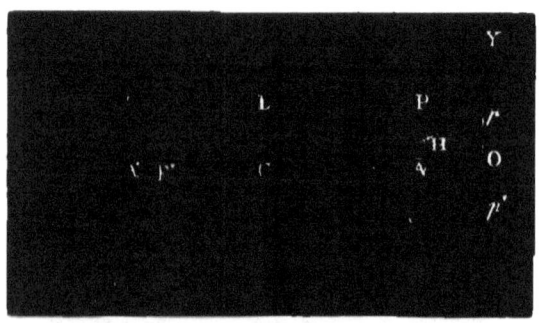

Fig. 2.

Let F be the focus, and Oy the directrix. Draw $FO \perp$ the directrix; divide FO internally at A, and externally at A', so that $FA : AO :: FA' : A'O$ in the given ratio, $\epsilon : 1$; then A and A' are the vertices of the curve, and AA' the axis. (*Def.*)

On the directrix take any point p; join Fp; draw AH and $A'H' \perp AA'$, meeting pF in H and H'; on HH' as diameter describe a circle; through p draw $pPP' \parallel$ the axis cutting the circle in P, P'; join PF, $P'F$.

Then since AH is $\parallel Op$; $\therefore HF : Hp :: AF : AO$;

also since $A'H'$ is $\parallel Op$; $\therefore H'F : H'p :: A'F : A'O$;

$\therefore H'F : H'p :: HF : Hp$.

Hence $PF : Pp = FH : Hp$ (Lemma.)

$= FA : AO = \epsilon : 1$.

(2 VI. Euclid.)

Also $P'F : P'p = FH : Hp$ (Lemma.)

$= FA : AO = \epsilon : 1$;

(2 VI. Euclid.)

$\therefore P$ and P' are points on the curve.

In like manner, by taking other points on the directrix, any number of points on the curve may be determined.

Cor. 1.—If Op' be taken equal to Op, and another point Q found in a similar manner to P, it is obvious that Qp' will be equal to Pp; therefore corresponding to any point P, at a perpendicular distance from the axis = Op, there is another point Q on the other side of the axis at a distance from it = Op, and also at the same distance as P from the directrix; hence the curve is symmetrical with regard to the axis.

Cor. 2.—Bisect AA' in C; through C draw $LCS \perp AA'$ and PP', meeting HH' in S, and PP' in L.

Then since AH, CS, and $A'H'$ are parallel;

$$\therefore SH = SH'.$$

Hence S is the centre of the circle $H'P'PH$;

and $\therefore LP = LP'$. (3 III. Euclid.)

Therefore corresponding to any point P on the curve there is another point P' on the other side CL, situated in precisely the same manner with regard to it as P; hence the curve is symmetrical also with regard to CL.

If, therefore, CO' be measured off = CO, and $CF' = CF$, and $O'y'$ be drawn $\perp O'C$, the curve could be equally well described with F' as focus, and $O'y'$ directrix.

Cor. 3. $PF : Pp = AF : AO = A'F : A'O$;
$\therefore PF : Pp = AF + A'F : AO + A'O$
$= 2CA : 2CO$;
$\therefore PF : Pp = CA : CO$,

that is, the distance of any point on the curve from the focus is to its distance from the directrix as $CA : CO$.

Cor. 4. $A'F : A'O = AF : AO$;
$\therefore A'F + AF : A'F - AF = A'O + AO : A'O - AO$;
$2CA : 2CF = 2CO : 2CA$;
$\therefore CF \times CO = CA^2$.

Cor. 5.—From the symmetry of the curve it is evident that any chord drawn through C is bisected at that point. From this property the point C is called the *centre* of the curve.

Def.—The chord drawn through the centre at right-angles to the Axis, or *Axis major*, as that portion of the axis intercepted between the vertices is frequently termed, is called the *Axis minor*.

Proposition II.

The sum of the distances of any point P on an ellipse from the foci is constant and equal to the axis-major.

Fig. 3.

$$PF : Pp = CA : CO.$$
(*Cor.* 3, Prop. i.)

Also $$PF' : Pp' = CA : CO;$$

$$\therefore PF + PF' : Pp + Pp' = CA : CO.$$

But $Pp + Pp' = 2CO$; $\therefore PF + PF' = 2CA = AA'$.

Cor. 1.—By the help of the preceding theorem we can describe an ellipse mechanically.

If the extremities of a thread, equal in length to AA', be fastened to the fixed points F, F', it is obvious that a pencil, moved about so as to keep the thread always stretched, will describe an ellipse whose foci are F and F'.

The Ellipse.

Cor. 2.—The distance of the extremity of the axis minor from either focus is equal to the semiaxis major. Let CB be the semiaxis minor; join BF, BF'.

Then the $\triangle^s$ BCF and BCF' are equal;

(4 I. Euclid.)

$$\therefore BF = BF' = \tfrac{1}{2}(BF + BF') = \tfrac{1}{2}AA' = CA.$$

Cor. 3.—$BC^2 = BF^2 - FC^2$

$$= CA^2 - CF^2 = A'F \times FA.$$

Cor. 4.—The sum of the distances of any point from the foci of an ellipse is greater or less than the axis major, according as the point is without or within the ellipse.

Fig. 4.

First let the point Q be without the ellipse; join QF, QF', and let QF' meet the ellipse in P; join PF.

Then $\qquad QF + QP > PF$;

$\qquad \therefore QF + QF' > PF + PF'$

$\qquad\qquad\qquad > AA'.$

Next, let Q' be within the ellipse; join $Q'F$, $Q'F'$, and let $F'Q'$ produced meet the ellipse in P; join PF.

Then $\qquad Q'F < Q'P + PF$;

$\qquad \therefore Q'F + Q'F' < PF' + PF$

$\qquad\qquad\qquad < AA'.$

Cor. 5.—Conversely, a point will be within or without an ellipse according as the sum of its distances from the foci is less than or greater than the axis major.

Proposition III.

1°. The line which bisects the external angle formed by drawing lines from any point on an ellipse to the foci falls wholly without the curve.

2°. Any other line drawn through the point will cut the curve.

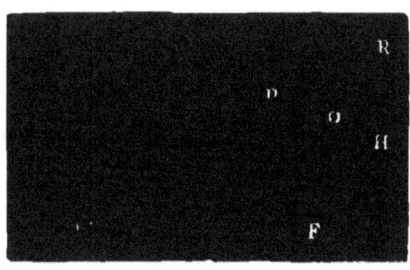

Fig. 5.

1°. Let PH bisect the angle between PF and PF' produced.

Let Q be any point on PH.

Cut off $PR = PF$.

Join QF, QF', QR.

Then the $\triangle^s\ QFP$ and QRP are equal; (4 I. Euclid.)

$$\therefore QR = QF;$$

but $\quad\quad F'Q + QR > F'R;\quad\quad$ (20 I. Euclid.)

$$\therefore F'Q + QF > F'P + PF$$
$$> \text{axis major};\quad\quad \text{(Prop. II.)}$$

$\therefore Q$ is a point without the curve.

(*Cor.* 5, Prop. II.)

CHAP. II.]   *The Ellipse.*   53

2°. Let PK be any line other than the bisector of the angle FPR.

Draw PL, making the $\angle KPL = \angle KPF$; cut off $PL = PF$; join $F'L$, and where it cuts PK join with F.

Fig. 6.

Then the $\triangle^s FPQ$ and LPQ are equal; (4 I. Euclid.)

$$\therefore QF = QL;$$

but $\qquad F'P + PL > F'L;$ (20 I. Euclid.)

$$\therefore F'P + PF > F'Q + QF;$$

$$\therefore AA' > F'Q + QF;$$

∴ the point Q is within the ellipse. (*Cor.* 5, Prop. II.)

Therefore the line PK will cut the curve.

Cor. 1.—Hence the line which bisects the external angle between lines drawn from any point on an ellipse to the foci is a tangent to the curve.

Cor. 2.—The tangent at the vertex is perpendicular to the axis, and hence ordinates to the axis are also perpendicular to the axis.

Cor. 3.—The normal to an ellipse at any point bisects the angle between the focal radii vectores drawn to the point.

Proposition IV.

The locus of the foot of the perpendicular let fall from either focus on any tangent to an ellipse is a circle described on the axis major as diameter.

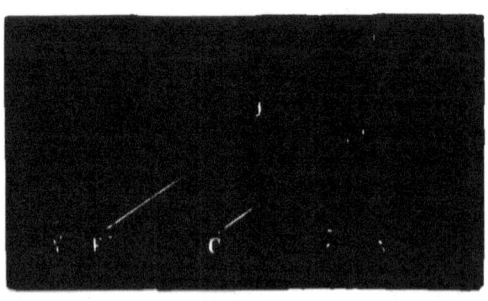

Fig. 7.

Let PH be any tangent, and FT a perpendicular let fall on it from the focus.

Produce $F'P$ to meet FT produced in R; join TC.

Then the $\triangle^s PFT$ and PRT are equal;
(26 I. Euclid.)

$$\therefore FT = TR.$$

Now in $\triangle F'FR$, $F'F$ is bisected in C, and FR in T.

$$\therefore CT \text{ is } \parallel F'R, \quad \text{(2 VI. Euclid.)}$$

and
$$CT = \tfrac{1}{2} F'R \quad \text{(Similar } \triangle^s.)$$
$$= \tfrac{1}{2}(F'P + FP)$$
$$= \tfrac{1}{2} AA'$$
$$= CA.$$

Hence the locus of T is a circle described with the centre C, and radius CA.

This circle is called *the Auxiliary circle* of the ellipse.

Cor. 1.—Conversely the right line drawn from either focus to the adjacent point of intersection of any tangent with the auxiliary circle is perpendicular to the tangent.

Cor. 2.—If the vertex of a right angle FTP move along a fixed circle, while one leg passes through a fixed point F within that circle, the other leg will always touch an ellipse.

Cor. 3.—By the aid of this Prop. we can draw a pair of tangents to an ellipse from any point Q without it.

On the line joining the given point Q, with either focus as diameter, describe a circle; the lines joining Q with the points of intersection of this circle with the auxiliary circle will be the required tangents.

NOTE.—The circle described on QF as diameter will always cut the auxiliary circle.

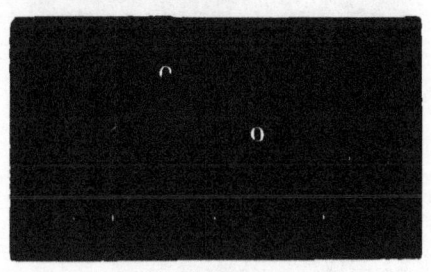

Fig. 8.

For, bisect QF in O; join OC, QF'.

Then, since $\quad QF + QF' > AA'$; (*Cor.* 4, Prop. III.)

$$\therefore OF + OC > CA;$$

$$\therefore OC > CA - OF.$$

That is, the distance between the centres of the circles is greater than the difference of their radii; ∴ they will intersect.

Cor. 4.—The right line drawn from the centre parallel to either focal radius vector, to meet the tangent, is equal to the semiaxis major.

Cor. 5.—The circle described on any focal radius vector as diameter will touch the auxiliary circle.

Proposition V.

The rectangle under the focal perpendiculars on any tangent to an ellipse is equal to the square of the semi-axis minor.

Fig. 9.

Let FT, $F'T'$, be perpendiculars let fall from the foci on the tangent at any point P.

Join TC, and produce it to meet $T'F'$ produced in S.

Then the $\triangle^s\ F'CS$ and FCT are equal; (26 I. Euclid.)

$\therefore F'S = FT$, and $CS = CT = CA$; (Prop. IV.)

$\therefore S$ is a point on the auxiliary circle;

$\therefore A'F' \times F'A = SF' \times F'T'$ (35 III Euclid.)

$= FT \times F'T'$;

$\therefore A'C^2 - CF'^2 = FT \times F'T'$, (5 II. Euclid.)

or $\qquad BC^2 = FT \times F'T'$. (*Cor.* 3, Prop. II.)

Proposition VI.

The locus of the intersection of the tangents to an ellipse which cut at right angles is a circle.

Fig. 10.

Let QP, QP', be tangents, which cut at right angles. Describe the auxiliary circle cutting the tangent QP in T and T', and the tangent QP' in H and H'; join FH, FT, $F'H'$, $F'T'$; draw QK touching the auxiliary circle; join CK, CQ.

Then FT and $F'T'$ are both $\perp QP$, and $\therefore \parallel QP'$.
(Cor. 1, Prop. iv.)

Similarly, $F'H'$ and FH are both $\perp QP'$, and $\therefore \parallel QP$;
(Cor. 1, Prop. iv.)

hence $\qquad QT = FH$, and $QT' = F'H'$.

Now $\qquad CQ^2 = CK^2 + QK^2$

$\qquad\qquad\quad = CA^2 + QT \times QT'$ (36 III. Euclid.)

$\qquad\qquad\quad = CA^2 + FH \times F'H'$

$\qquad\qquad\quad = CA^2 + CB^2$ (Prop. v.)

$\qquad\qquad\quad = AB^2$;

$\therefore CQ = AB$.

Hence the locus of Q is a circle described with the centre C, and radius $= AB$.

This circle is called *the Director Circle* of the ellipse.

Proposition VII.

If PN be a normal to an ellipse at any point P, and PM an ordinate to the axis, then
$$CM : CN = CA^2 : CF^2.$$

Fig. 11.

Join PF, PF'; draw $p'Pp$ ∥ the axis major, meeting the directrices in p', p.

Since (*Cor.* 3, Prop. III.) PN bisects the ∠ $F'PF$;

$$\therefore NF' : NF = PF' : PF \quad (3 \text{ VI. Euclid.})$$
$$= Pp' : Pp.$$

Comp. and div.

$$NF' + NF : NF' - NF = Pp' + Pp : Pp' - Pp,$$

or
$$2CF : 2CN = 2CO : 2CM;$$
$$\therefore CM : CN = CO : CF$$
$$= CO \times CF : CF^2.$$

Hence $\quad CM : CN = CA^2 : CF^2.$

(*Cor.* 4, Prop. I.)

Cor. 1.—$CM : MN = CA^2 : CB^2.$

For $\quad CM : CM - CN = CA^2 : CA^2 - CF^2,$

or $\quad CM : MN = CA^2 : CB^2.$

[CHAP. II.] *The Ellipse.* 59

Cor. 2. $F'P : PF = F'N : NF$

$\therefore F'P + PF : PF = F'N + NF : NF$,

or $CA : CF = PF : NF$.

Similarly, $CA : CF = PF' : NF'$;

$\therefore CA^2 : CF^2 :: PF \times PF' : NF \times NF'$

$\therefore CA^2 : CA^2 - CF^2 :: PF \times PF' : PF \times PF' - NF \times NF'$

$CA^2 : CB^2 :: PF \times PF' : PN^2$.

Proposition VIII.

If a tangent at any point P of an ellipse be produced to meet the axis in T, and the ordinate PM be drawn, then CA will be a mean proportional between CM and CT.

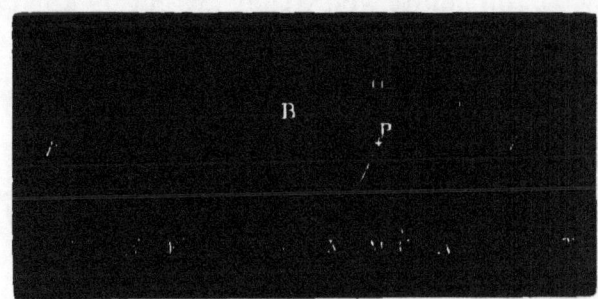

Fig. 12.

Through P draw $p'Pp \parallel$ the axis; join PF, PF'.
Then since PT is the bisector of the $\angle FPH$;

$\therefore F'T : TF = F'P : PF$ (*Cor.* 3, VI. Euclid.)

$= p'P : Pp$.

Comp. and div.

$F'T + TF : F'T - TF = Pp' + Pp : Pp' - Pp$,

or $2CT : 2CF = 2CO : 2CM$;

$\therefore CT \times CM = CF \times CO$

$= CA^2$. (*Cor.* 4, Prop. I.)

Cor. 1.—If the auxiliary circle be described, and PM produced to meet it in Q, then QT will be a tangent to the circle.

For $\quad CT : CA = CA : CM$;

$\therefore CT : CQ = CQ : CM$;

$\therefore$ the $\triangle^s CQT$ and CMQ are similar;
(6 VI. Euclid.)

$\therefore \angle CQT = \angle CMQ = 90°$;

$\therefore QT$ is a tangent to the circle.

Cor. 2.—Tangents to the ellipse and auxiliary circle, respectively, at the points where they are cut by any ordinate to the axis, will meet the axis produced in the same point.

Cor. 3.—Corresponding ordinates PM, QM, of the ellipse and auxiliary circle are in a constant ratio.

Since the $\triangle NPT$ is right-angled, and $PM \perp NT$;

$\therefore NM \times MT = PM^2$.

Also, since $\triangle CQT$ is right-angled, and $QM \perp CT$;

$\therefore CM \times MT = QM^2$.

Hence $\quad PM^2 : QM^2 = NM : CM$

$\qquad\qquad\qquad = CB^2 : CA^2$.
(*Cor.* 1, Prop. vii.)

$\therefore PM : QM = CB : CA$.

NOTE.—The angle QCA is called *the excentric angle* of the point P.

The Ellipse.

Cor. 4.—The ellipse is concave towards the axis major.

Fig. 13.

$$QM : PM = AC : BC = qm : pm. \quad (Cor.\ 3.)$$

But $\quad QM : PM = sm : rm ; \qquad$ (Similar $\triangle^s$.)

$$\therefore qm : pm = sm : rm.$$

But qm is $> sm$ by property of the circle;

$$\therefore pm \text{ is } > rm.$$

That is, if a perpendicular be drawn to the axis major, between the vertex A and any point P on the curve, the segment intercepted on it between the curve and the axis major is greater than the segment intercepted on it between the line AP and the axis major; and, therefore, the curve is concave towards the axis major.

Proposition IX.

The rectangle under the segments of the axis made by any ordinate is to the square of the ordinate in a constant ratio. (See Fig. 12.)

$$PM^2 : QM^2 = CB^2 : CA^2, \quad \text{(Prop. VIII.)}$$

but $\quad QM^2 = A'M \times MA ; \quad$ (35 III. Euclid.)

$$\therefore PM^2 : A'M \times MA = CB^2 : CA^2.$$

Cor. 1.—The Latus rectum is a third proportional to the axis major and axis minor. (See Fig. 12.)

For $\left(\dfrac{L}{2}\right)^2 : AF \times FA' = CB^2 : CA^2$

$\left(\dfrac{L}{2}\right)^2 : CB^2 = CB^2 : CA$; (*Cor.* 3, Prop. II.)

hence $CA : CB = CB : \dfrac{L}{2}$.

Cor. 2.—If from the foot of any ordinate a line be drawn parallel to the line joining the extremities of the axes, the square of the ordinate will be equal to the rectangle under the segments of the axis minor.

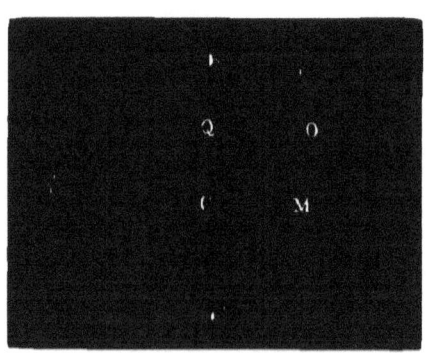

Fig. 14.

$CA^2 : CM^2 = CB^2 : CQ^2$;   (Parallel lines.)

$\therefore CA^2 : CA^2 - CM^2 = CB^2 : CB^2 - CQ^2$;

$\therefore CA^2 : AM \times MA' = CB^2 : BQ \times QB'$;

$\therefore CB^2 : PM^2 = CB^2 : BQ \times QB'$;

$PM^2 = BQ \times QB'$.

CHAP. II.]   The Ellipse.   63

Cor. 3.—If PN be a normal at point P, meeting the axis major in N, and NH a perpendicular let fall on the focal radius vector, then PH is equal to half the Latus rectum.

Fig. 15.

For $\quad FT : FP = PH : PN.$ (Similar $\triangle^s$.)

Also $\quad F'T' : F'P = PH : PN;$ (Similar $\triangle^s$.)

$$\therefore FT \times F'T' : FP \times F'P = PH^2 : PN^2$$

$$CB^2 : FP \times F'P = PH^2 : PN^2;$$ (Prop. V.)

but $\quad CA^2 : FP \times F'P = CB^2 : PN^2;$ (*Cor.* 2, Prop. VII.)

$$\therefore CA : CB = CB : PH;$$

$$\therefore PH = \tfrac{1}{2} \text{ Latus rectum.} \quad (Cor. 1, \text{Prop. IX.})$$

Cor. 4.—The normal and the focal perpendiculars on the tangent at any point are in Harmonic proportion.

Let the tangent $T'T$, when produced, meet the axis in Q.

Then $\quad T'Q : QT = F'T' : FT.$ (Similar $\triangle^s$.)

Also $\quad F'T' : FT = T'P : PT;$ (Similar $\triangle^s$.)

$$\therefore T'Q : QT = T'P : PT;$$

$\therefore T'Q, PQ,$ and TQ; and $\therefore F'T', NP,$ and $FT,$ are in Harmonic proportion.

Proposition X.

The locus of the intersection of any tangent to an ellipse, with the line drawn through the focus perpendicular to the radius vector drawn from the focus to the point of contact of the tangent, is the directrix.

Fig. 16.

Let PT be any tangent, $FT \perp FP$.

Draw $TH \perp F'P$ produced, and $TO \perp F'F$ produced; join TF'.

The $\triangle^s\ TFP$ and THP are equal; (26 I.Euclid.)

$$\therefore FT = TH \text{ and } PF = PH;$$

$$\therefore F'H = F'P + PF = AA'.$$

Now $F'H^2 = F'T^2 - TH^2 = F'T^2 - TF^2$

$$= F'O^2 - OF^2$$

$$= 2CO \times 2CF;$$

$$\therefore AA'^2 = 4CO \times CF, \text{ or } CA^2 = CO \times CF.$$

Hence OT is the directrix. (*Cor.* 4, Prop. I.)

Cor. 1.—Conversely, if from any point on the directrix a tangent be drawn to an ellipse, the line joining that point to the focus is perpendicular to the radius vector drawn from the focus to the point of contact of the tangent.

Cor. 2.—Tangents at the extremities of a focal chord intersect on the directrix.

Proposition XI.

If any chord PP' of an ellipse cut the directrix in D, and if F be the focus corresponding to the directrix on which D is situated, then FD is the external bisector of the angle PFP'.

Fig. 17.

Produce $P'F$ to Q; draw Pp, $P'p'$ ⊥ the directrix.

Then $\quad P'F : PF = P'p' : Pp$

$\qquad\qquad\qquad = P'D : PD;\quad$ (Similar △'.)

∴ FD bisects the ∠ PFG. (Prop. B., VI. Euclid.)

Cor. 1.—Hence, being given one focus F, and three points P, P', P'', on an ellipse, we can find the directrix and the axes.

For, draw FD bisecting the ∠ between PF and $P'F$ produced: the point where this line meets $P'P$ produced will be one point on the directrix. Similarly, by bisecting the external ∠ between PF and $P''F$, another point on the directrix may be found, and hence the directrix itself.

Also, since the eccentricity is the ratio of PF to the perpendicular distance of P from the directrix, the axes may be found by Prop. I., and Cor.

If a tangent to a curve be defined as *The line joining two indefinitely near points on the curve*, it will follow immediately from this Prop. that *the right line drawn from the focus to the point of intersection of any tangent with the directrix is perpendicular to the radius vector drawn form the focus to the point of contact;* also that *the tangent makes equal ∠^{s} with the radii vectores drawn from the foci to the point of contact.*

See Fig. 17.

For, when P is indefinitely near to P', PD becomes a tangent at the point P; and the ∠ PFP' is indefinitely small;

$$\therefore PFQ = 180°;$$

but $\quad\quad\quad\quad ∠ PFD = \tfrac{1}{2} ∠ PFQ;$

$$\therefore ∠ PFD = 90°.$$

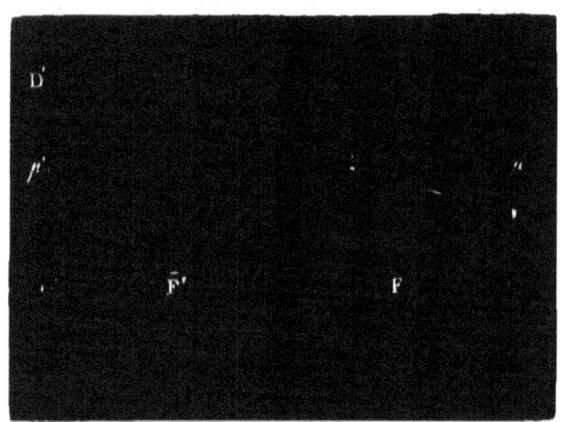

Fig. 18.

Again, $\quad PF : PF' = Pp : Pp' = PD : PD',$
$\quad\quad\quad\quad\quad\quad\quad\quad\quad\quad\quad\quad$ (2, VI. Euclid.)

and $\quad\quad ∠ PFD = ∠ PF'D'$, each being 90°.

Hence $\quad \triangle PFD$ is similar to $\triangle PF'D'$; (7, VI. Euclid.)

$$\therefore ∠ FPD = ∠ F'PD'.$$

Proposition XII.

If two fixed points P, P', on an ellipse be joined with a third variable point O, the segment pp', intercepted on either directrix by the produced chords, subtends a constant angle at the focus corresponding to that directrix.

Fig. 19.

Since (by Prop. XI.) Fp is the bisector of the external $\angle$ between FO and FP;

$$\therefore \angle OFp + \tfrac{1}{2} \angle OFP = 90°.$$

Also Fp' is the bisector of the external $\angle$ between FO and FP';

$$\therefore \angle OFp' + \tfrac{1}{2} \angle OFP' = 90°;$$

$$\therefore \angle OFp + \tfrac{1}{2} \angle OFP = \angle OFp' + \tfrac{1}{2} \angle OFP';$$

$$\therefore \angle OFp - \angle OFp' = \tfrac{1}{2}(\angle OFP' - \angle OFP),$$

or $\qquad \angle pFp' = \tfrac{1}{2} \angle PFP'$, and $\therefore$ constant.

Cor. 1.—Hence the anharmonic ratio of the pencil formed by joining four fixed points on an ellipse to any fifth variable point is constant.

For $\quad O \cdot PP'P''P''' = O \cdot pp'p''p''' = F \cdot pp'p''p'''.$

Proposition XIII.

The line joining the focus to the intersection of two tangents to an ellipse bisect the angle which the points of contact subtend at the focus.

Fig. 20.

Let QP, QP', be tangents; join PF, PF', $P'F'$, $P'F$.

Produce FP till $PR = PF'$, and FP' till $P'R' = P'F'$; join QR, QR'.

The $\triangle^s$ $QF'P'$ and $QR'P'$ are equal; $\therefore QR' = QF'$.
(4, I. Euclid.)

The $\triangle^s$ QPF' and QPR are equal; $\therefore QR = QF'$;
(4, I. Euclid.)

hence $QR = QR$.

But $FR' = FP' + P'F' = FP + PF' = FR$;

hence the $\triangle^s FQR$ and FQR' are equal; (8, I. Euclid.)

$\therefore \angle QFR = \angle QFR'$.

Cor. 1.—It is obvious from the above demonstration that
$$\angle QF'P' = \angle QR'P' = \angle QRP = \angle QF'P.$$

CHAP. II.] The Ellipse. 69

Note.—If the ordinate at any point P on an ellipse be produced to meet the Auxiliary circle in Q, then

$\tan \tfrac{1}{2} PF'C = \sqrt{\dfrac{1-e}{1+e}} \cdot \tan \tfrac{1}{2} QCF$; and $\tan \tfrac{1}{2} PFA = \sqrt{\dfrac{1+e}{1-e}} \tan \tfrac{1}{2} QCF$.

Fig. 21.

Draw the tangent at the vertex meeting the tangents at P and Q in H and K respectively; join KC, HF', HF. Let $CA = a$.

Then $\angle HF'C = \tfrac{1}{2} \angle PF'C$ and $\angle HFA = \tfrac{1}{2} \angle PFA$. (Prop. XIII.)
Also $\angle KCA = \tfrac{1}{2} \angle QCF$.
Now $\tan \tfrac{1}{2} PF'C = \tan HF'C = \dfrac{HA}{AF'}$,

$\tan \tfrac{1}{2} PFA = \tan HFA = \dfrac{HA}{AF}$,

$\tan \tfrac{1}{2} QCF = \tan KCA = \dfrac{KA}{AC}$;

$\therefore \dfrac{\tan \tfrac{1}{2} PF'C}{\tan \tfrac{1}{2} QCF} = \dfrac{HA \times AC}{KA \times AF'} = \dfrac{PM}{QM} \cdot \dfrac{AC}{AF'} = \dfrac{CB}{CA} \cdot \dfrac{AC}{AF'}$ (Cor. 3, Prop. VIII.)

$= \dfrac{CB}{AF'} = \dfrac{\sqrt{CA^2 - CF^2}}{CA + CF} = \sqrt{\dfrac{CA - CF}{CA + CF}} = \sqrt{\dfrac{a - ae}{a + ae}}$;

$\therefore \tan \tfrac{1}{2} PF'C = \sqrt{\dfrac{1-e}{1+e}} \cdot \tan \tfrac{1}{2} QCF$.

Again, $\dfrac{\tan \tfrac{1}{2} PFA}{\tan \tfrac{1}{2} QCF} = \dfrac{HA \times AC}{KA \times AF} = \dfrac{PM}{QM} \cdot \dfrac{AC}{AF} = \dfrac{CB}{CA} \cdot \dfrac{AC}{AF}$ (Cor. 3, Prop. VIII.)

$= \dfrac{CB}{AF} = \dfrac{\sqrt{a^2 - a^2 e^2}}{a - ae} = \sqrt{\dfrac{1+e}{1-e}}$;

$\therefore \tan \tfrac{1}{2} PFA = \sqrt{\dfrac{1+e}{1-e}} \cdot \tan \tfrac{1}{2} QCF$.

Cor. 1. $\dfrac{\tan \tfrac{1}{2} PF'C}{\tan \tfrac{1}{2} PFA} = \dfrac{1-e}{1+e}$.

70 · The Ellipse. · [CHAP. II.

Cor. 2.—If a quadrilateral be circumscribed to an ellipse, its opposite sides subtend supplementary angles at either focus.

Cor. 3.—By the aid of this Prop., also, we can draw a pair of tangents to an ellipse from any point without it.

With Q as centre and QF' as radius, describe a circle; with F as centre and a radius = the axis major, describe a circle; the lines joining the points of intersection of these circles with F will determine the points of contact of the tangents.

Proposition XIV.

The angle at either focus, subtended by the segment intercepted on a variable tangent to an ellipse by two fixed tangents, is constant.

Fig. 22.

Let QP, QP', be the fixed tangents, RS the variable tangent; join FP', FR, FT, FS, FP.

Then $\angle TFS = \frac{1}{2} \angle TFP$, also $\angle TFR = \frac{1}{2} \angle TFP'$;

(Prop. XIII.)

∴ $\angle RFS = \frac{1}{2} \angle P'FP$, and ∴ constant.

Cor. 1.—The anharmonic ratio of the four points in which a variable tangent to an ellipse is cut by four fixed tangents is constant.

For the segments intercepted on the variable tangent subtend constant angles at the focus.

The Ellipse.

Proposition XV.

If a tangent at the extremity of any diameter CP meet the axis produced in T, then the area of the $\triangle\, CPT$ thus formed will be equal to the area of the $\triangle\, CAH$, formed by producing the diameter CP to meet the tangent at the vertex A.

Fig. 23.

Draw the ordinate PM; join AP, HT.

Then $\quad\quad TC : CA = CA : CM \quad\quad$ (Prop. VIII.)

$\quad\quad\quad\quad\quad\quad\quad = CH : CP; \quad\quad$ (Parallel lines.)

$\therefore\ PA$ is $\parallel HT$; $\quad\quad$ (2, VI. Euclid.)

$\therefore\ \triangle\, PHA = \triangle\, PTA$; $\quad\quad$ (37, I. Euclid.)

$\therefore\ \triangle\, CHA = \triangle\, CPT$.

Cor. 1.— $\triangle\, PMT$ = trap. $PMAH$.

Def.—The chords which join the extremities of any diameter to any point on the curve are called *supplemental chords*.

The axis major is sometimes called the *Transverse axis*, and the axis minor the *Conjugate axis*.

NOTE.—Throughout the following Props. CA denotes the semi-axis major, and CB the semi-axis minor.

Proposition XVI.

If an ordinate, drawn to the axis from any point P on an ellipse, be produced to meet the auxiliary circle in Q, and CQ' be drawn perpendicular to CQ, the perpendicular drawn from Q' to the axis will meet the ellipse in a point P', through which, if a diameter be drawn, it will be parallel to the tangent at P.

Fig. 24.

The tangent to the circle at Q will intersect the tangent to the ellipse at P on the axis. (*Cor.* 2, Prop. VIII.)

Then, since QT and $Q'C$ are both $\perp CQ$; ∴ they are parallel.

Hence $\triangle^s\ QMT$ and $Q'TM'$ are similar;

∴ $MT : M'C = QM : Q'M'$ (Similar $\triangle^s$.)

$= PM : P'M'$. (*Cor.* 3, Prop. VIII.)

Hence $\triangle^s\ PMT$ and $P'M'C$ are similar;

∴ $P'C$ is $\parallel PT$.

Def.—The diameter which is parallel to the tangent at any point is said to be *conjugate* to the diameter which passes through that point.

Cor. 1.—If we draw tangents at P' and Q', it can be similarly shown that CP is parallel to the tangent at P': hence, if the diameter which passes through P' be conjugate to that through P; conversely, the diameter which passes through P will be conjugate to that through P'.

Cor. 2. $CM = Q'M'$, and $CM' = QM$.

Since $\angle M'Q'C + \angle M'CQ' = 90°$,

and also $\angle M'CQ' + \angle MCQ = 90°$;

$\therefore \angle M'Q'C = \angle MCQ$.

Hence the $\triangle^s M'Q'C$ and MCQ are similar;

but $CQ' = CQ$; $\therefore CM = QM'$, and $CM' = QM$.

Cor. 3. $CM^2 + CM'^2 = CA^2$.

For $CA^2 = CQ^2 = CM^2 + QM^2 = CM^2 + CM'^2$.   (*Cor.* 2.)

Cor. 4. $PM^2 + P'M'^2 = CB^2$.

Since $QM : Q'M' = PM : P'M'$; (*Cor.* 3, Prop. VIII.)

$\therefore QM^2 + Q'M'^2 : QM^2 = PM^2 + P'M'^2 : PM^2$.

Alt. $\therefore CA^2 : PM^2 + P'M'^2 = QM^2 : PM^2$ (*Cors.* 2 & 3.)

$= CA^2 : CB^2$;

(*Cor.* 3, Prop. VIII.)

$\therefore PM^2 + P'M'^2 = CB^2$.

Cor. 5.—The sum of the squares of any pair of semi-conjugate diameters is equal to the sum of the squares of the semi-axes.

For $CM^2 + CM'^2 = CA^2$,   (*Cor.* 2.)

and $PM^2 + P'M'^2 = CB^2$;   (*Cor.* 3.)

$\therefore CP^2 + CP'^2 = CA^2 + CB^2$.

Cor. 6. $PM : CM' = CB : CA :: P'M' : CM$.

For $PM : QM = CB : CA$; but $QM = CM'$;

$\therefore PM : CM' = CB : CA$.

Again $P'M' : Q'M' = CB : CA$; but $Q'M' = CM$;

$\therefore P'M' : CM = CB : CA$.

Cor. 7.—The $\triangle^s CPM$ and $CP'M'$ are equal in area.

For $PM : P'M = QM : Q'M' = CM' : CM$; (*Cor.* 2.)

$\therefore PM \times CM = P'M' \times CM'$.

Proposition XVII.

If QG be an ordinate drawn to any diameter Pp from any point Q on an ellipse, cutting the transverse axis in L, QS an ordinate drawn to the axis from the same point, and produced, if necessary, to meet the diameter Pp in K; AH a tangent at the vertex: then $\triangle QLS$ = trap. $KSAH$.

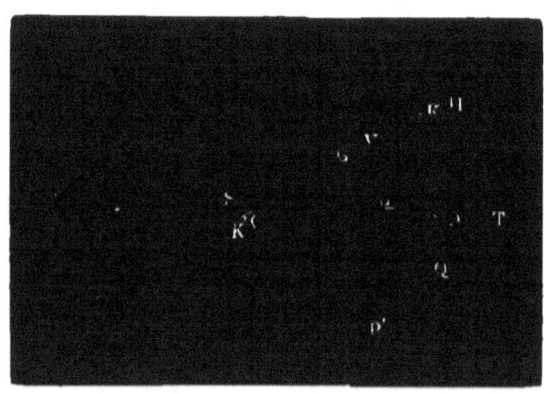

Fig. 25.

Draw the ordinate PM to the axis.

Then $CA : CS : CM = AH : SK : MP$;
(Similar $\triangle$'.)

$\therefore CA + CS : CA + CM = AH + SK : AH + MP$,

or $\quad A'S : A'M = AH + SK : AH + MP$;

$\therefore A'S \times AS : A'M \times AM = (AH + SK) AS :$
$(AH + MP) AM$;

$\therefore SQ^2 : PM^2$ = trap. $KSAH$: trap. $PMAH$;
(Prop. ix.)

$\therefore \triangle QSL : \triangle PMT$ = trap. $KSAH$: trap. $PMAH$;
(19, VI. Euclid.)

but $\quad \triangle PMT$ = trap. $PMAH$; (Cor. 1, Prop. xv.)

$\therefore \triangle QSL$ = trap. $KSAH$.

Cor. 1.— △ QGK = trap. $TPGL$.

For   △ QSL = trap. $KSAH$;

∴   △ QGK = trap. $GLAH$ = trap. $TPGL$.

Cor. 2.—In like manner it can be shown that

△ $GQ'K'$ = trap. $TPGL$;

∴ △ GQK = △ $GQ'K'$;

but they are also similar;

∴ $GQ = GQ'$.

Hence any diameter bisects all chords parallel to the tangents at its extremities.

Proposition XVIII.

If CP' be the semi-diameter conjugate to CP, QG any ordinate to CP; then $QG^2 : PG \times Gp = CP'^2 : CP^2$.

For   △ CPT : △ $CGL = CP^2 : CG^2$.

Division. △ CPT : trap. $TPGL = CP^2 : CP^2 - CG^2$,

or   △ $CP'V$: △ $QGK = CP^2 : PG \times Gp$;

(*Cor.* 1, Prop. XVIII.)

∴ $CP'^2 : GQ^2 = CP^2 : PG \times Gp$,

(19, VI. Euclid.)

or $QG^2 : PG \times Gp = CP'^2 : CP^2$.

Cor. 1.—Ordinates to any diameter at equal distances from the centre are equal.

For the rectangles are the same for equal distances from the centre.

Cor. 2.—The squares of the ordinates to any diameter are proportional to the rectangles under the segments of the diameter.

Proposition XIX.

If a normal at any point P of an ellipse meet the axis major in N, and the axis minor in N'; then
$$PN : CP' :: CB : CA :: CP' : PN',$$
CP' being the semi-diameter conjugate to CP.

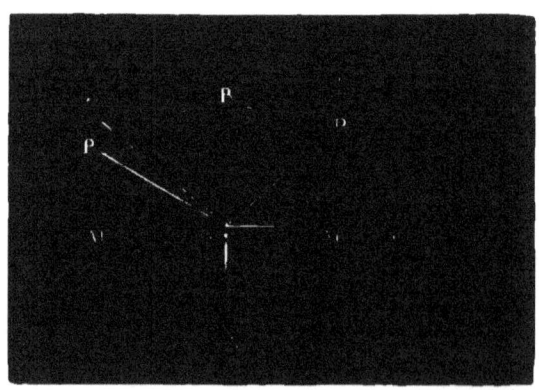

Fig. 26.

Since CP' is $\parallel TP$, the $\triangle P'CM'$ is similar to the $\triangle PMT$, and $\therefore$ similar to $\triangle NPM$.   (8, VI. Euclid.)

$$\therefore P'C : NP = CM' : PM$$
$$= QM : PM \text{ (Cor. 2, Prop. xvi.)}$$
$$= CA : CB. \text{ (Cor. 2, Prop. viii.)}$$

Also $\quad PN' : PN = CM : MN \quad$ (Similar $\triangle^s$.)
$$= CA^2 : CB^2 ; \quad \text{(Prop. vii.)}$$
$$\therefore PN' \times PN : PN^2 = CA^2 : CB^2$$
$$= P'C^2 : PN^2 ;$$
$$\therefore PN' \times PN = P'C^2 ;$$
$$\therefore PN' : P'C = P'C : PN$$
$$= CA : CB.$$

Cor. $\quad PN \times PN' = P'C^2.$

CHAP. II.] *The Ellipse.* 77

PROPOSITION XX.

The rectangle under the distances of the foci, from any point P on an ellipse, is equal to the square of the semi-diameter CP', conjugate to that passing through the point.

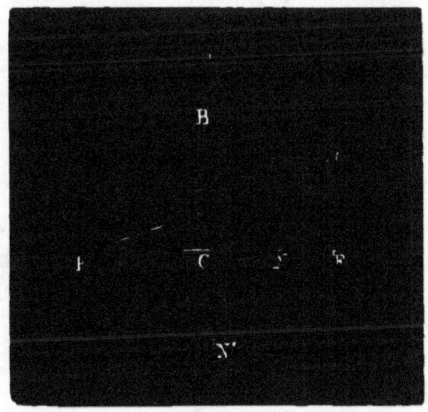

Fig. 27.

Circumscribe the $\triangle F'PF$ by a circle cutting the axis minor in N' and L; join PN' and $N'F$.

Since CL bisects FF', and is also $\perp$ it; $\therefore$ LN' is the diameter of the circle, and arc $F'N'$ = arc FN';

$$\therefore \angle F'PN' = \angle FPN'. \quad (26, \text{III. Euclid.})$$

Also $\angle PF'N = \angle PN'F$;

hence the $\triangle^s F'PN$ and $N'PF$ are similar;

$$\therefore F'P : PN = N'P : PF;$$

$$\therefore F'P \times PF = PN \times N'P$$

$$= CP'^2. \quad (Cor.\ 1,\ \text{Prop. XIX.})$$

Cor. 1.—$PN^2 = PN' \times PN - N'N \times NP$

(3, II. Euclid.)

$$= CP'^2 - F'N \times NF.$$

Otherwise thus:—

$$2CA = PF + PF';$$
$$\therefore 4CA^2 = PF^2 + PF'^2 + 2PF \times PF' \quad (4, \text{II. Euclid.})$$
$$= 2CP^2 + 2CF^2 + 2PF \times PF';$$
$$\therefore PF \times PF' = 2CA^2 - CF^2 - CP^2$$
$$= CA^2 + CB^2 - CP^2 \quad (Cor.\ 3,\ \text{Prop. II.})$$
$$= CP^2 + CP'^2 - CP^2 \quad (Cor.\ 3,\ \text{Prop. XVI.})$$
$$= CP'^2.$$

Cor. 2.—Since PN' bisects the $\angle FPF'$ it is a normal, and $\therefore PL$, which is $\perp PN'$, is a tangent. Hence the circle which passes through the foci, and any point P on the ellipse, passes also through the points in which the tangent and normal at P intersect the axis minor.

Cor. 3.—If K be the foot of the perpendicular let fall from N' on either PF' or PF produced,

then $\quad PK = \tfrac{1}{2}(PF + PF') = $ semi-axis major.

This follows immediately from the well-known properties of the triangle FPF', and its circumscribing circle. (See Geometrical Gymnasium, Exercise 179, Galbraith and Haughton's Manual of Euclid, books i., ii., iii.)

Cor. 4.—If K' be the foot of the perpendicular let fall from L on PF',

then $\quad F'K' = \tfrac{1}{2}(PF + PF') = $ semi-axis major.

Also if K'' be the foot of the perpendicular let fall from L on FP produced,

then $\quad FK'' = \tfrac{1}{2}(PF + PF') = $ semi-axis major.

(See as in *Cor.* 3.)

Cor. 5. $\quad CA^2 : CB^2 = PF' \times PF : PN^2$
$$(Cor.\ 2,\ \text{Prop. VII.})$$
$$= CP'^2 : PN^2;$$
$$\therefore CA : CB = CP' : PN.$$

Proposition XXI.

If CK be a perpendicular let fall from the centre on a tangent to an ellipse at any point P, and CP' the semi-diameter conjugate to CP, then $CK : CA :: CB : CP'$.

Fig. 28.

Draw $FT, F'T' \perp$ the tangent at P.
Since the $\triangle^s FPT$ and $F'PT'$ are similar;
$$\therefore FP : F'P = FT : F'T';$$
$$= FP + F'P : FT + F'T'$$
$$= CA : CK;$$
$$\therefore FP \times F'P : FT \times F'T' = CA^2 : CK^2;$$
but $FP \times F'P = CP'^2,$ (Prop. xx.)
and $FT \times F'T' = BC^2;$ (Prop. v.)
$$\therefore CP'^2 : BC^2 = CA^2 : CK^2;$$
$$\therefore CP' : BC = CA : CK.$$

Cor.—The area of the triangle formed by joining the extremities of any pair of conjugate semi-diameters is constant.

For $CP' \times CK = CB \times CA$;
$\therefore$ area of $\triangle P'CP =$ area of $\triangle BCA$, and $\therefore$ constant.

Proposition XXII.

If two chords of an ellipse intersect one another, the rectangles under the segments are proportional to the squares of the parallel semi-diameters.

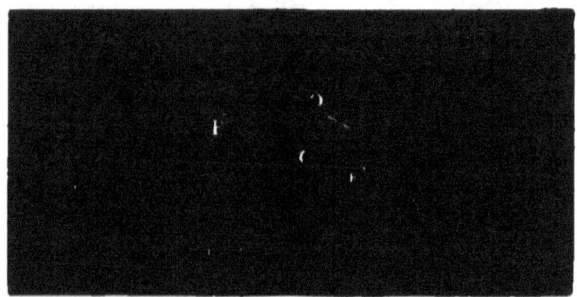

Fig. 29.

Let PP' be any chord drawn through O, and CR the parallel semi-diameter. Draw the ordinates PM, $P'M'$, RN to the axis, and produce them to meet the auxiliary circle in Q, Q', S.

Then $PM : P'M' = QM : Q'M'$;   (*Cor.* 2, *Prop.* VIII.)

∴ PP' and QQ', if produced, will meet the axis produced in the same point T.

Through O draw $O'OL \perp$ the axis.

Again $NC : MT = RN : P'M'$ (Similar △^{s}.)
          $= SN : Q'M'$;
                              (*Cor.* 3, *Prop.* VIII.)

∴ SC is $\parallel Q'T$; hence △ SCR is similar to △ $Q'TP'$.

Now $OP' : O'Q' = P'T : Q'T$,   (Parallel lines.)
and $OP : O'Q = P'T : Q'T$;
∴ $OP \times OP' : O'Q \times O'Q' = P'T^2 : Q'T^2$
                              $= RC^2 : CS^2$.

Alt. $OP \times OP' : CR^2 = O'Q \times O'Q' : CS^2$.

CHAP. II.] *The Ellipse.* 81

Similarly, if any other chord pp' be drawn through O, and the ordinates pm, $p'm'$ produced to meet the auxiliary circle in q, q': then, since

$$OL : O'L = CB : CA$$
$$= pm : qm = p'm' : q'm';$$
(*Cor.* 3, Prop. VIII.)

∴ the chord qq' will pass through O';
and if Cr be the semi-diameter parallel to pp',

then $Op \times Op' : Cr^2 = O'q \times O'q' : CA^2$.

But $O'Q \times O'Q' = O'q \times O'q'$; (35, III. Euclid.)

hence $OP \times OP' : Op \times Op' = CR^2 : Cr^2$.

(*See Note, page* 83.)

Otherwise thus :—

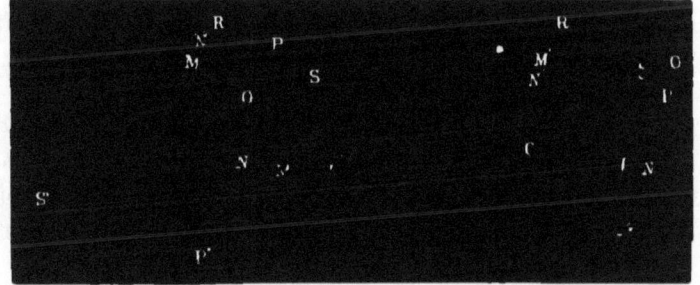

Fig. 30. Fig. 31.

Let PP' be any chord drawn through the point O, and CR the parallel semi-diameter, CR' the semi-diameter conjugate to CR; draw the diameter through O, meeting the curve in S; draw SM an ordinate to CR', and SM', PN' ordinates to CR.

Then $RC^2 - CN'^2 : RC^2 - CM'^2 = PN'^2 : SM'^2$;
(*Cor.* 2, Prop. XVIII.)

∴ $RC^2 - PN^2 : RC^2 - SM^2 = CN^2 : CM^2$
$= ON^2 : SM^2$.

Alt. $RC^2 - PN^2 : ON^2 = RC^2 - SM^2 : SM^2$.

Comp. $RC^2 - PN^2 + ON^2 : ON^2 = RC^2 : SM^2$.

G

82   *The Ellipse.*   [CHAP. II.

Alt. $RC^2 - PO \times OP' : RC^2 = ON^2 : SM^2$
$= OC^2 : CS^2.$

Div. $PO \times OP' : RC^2 = CS^2 - OC^2 : CS^2$
$= SO \times OS' : CS^2.$

If pp' be any other chord drawn through O, and Cr the parallel semi-diameter; then

$$pO \times Op' : Cr^2 = SO \times OS' : CS^2.$$

Hence $PO \times OP' : pO \times Op' = CR^2 : Cr^2.$

Cor. 1.—The rectangle under the segments of any diameter made by an ordinate is to the square of the ordinate as the square of the semi-diameter is to the square of its semi-conjugate.

This follows by supposing one of the chords to pass through the centre, and the other parallel to the tangent at its extremity.

Cor. 2.—If O be without the ellipse, and P, P' coincide, also p, p'; then the tangents to an ellipse from any point without it are proportional to the parallel semi-diameters.

Cor. 3.—If two parallel tangents OP, $O'P'$, be met by any third tangent, OO'; then $OP : O'P' = OQ : O'Q$.

Fig. 32.

For $OP : OQ = CR : Cr = O'P' : O'Q.$ (*Cor.* 2.)

CHAP. II.] *The Ellipse.* 83

Cor. 4.—If from a point without an ellipse a secant and also a tangent be drawn, the rectangle under the whole secant and its external segment is to the square of the tangent as the squares of the parallel semi-diameters.

NOTE.—If for O, O', L, we read o, o', l, the demonstration, page 80, will apply to the case when the point in which the chords intersect lies without the ellipse.

PROPOSITION XXIII.

If from a point without an ellipse two tangents OP, OR be drawn, any line AQ' drawn parallel to either will be cut by the curve and chord of contact of the tangents in Geometric proportion.

Fig. 33.

Draw the semi-diameters CE and $CD \parallel OP$ and OR respectively.

Then $\quad AQ \times AQ' : AP^2 = CD^2 : CE^2$
$\qquad\qquad\qquad$ (*Cor.* 4, Prop. XXII.)
$\qquad\qquad = OR^2 : OP^2$
$\qquad\qquad\qquad$ (*Cor.* 2, Prop. XXII.)
$\qquad\qquad = AB^2 : AP^2$;
$\qquad\qquad\qquad$ (Parallel lines.)
$\therefore AQ \times AQ' = AB^2.$

Cor.—If AQ' be a tangent: then $AQ = AB$.

That is, if two parallel tangents to an ellipse be cut by any third tangent, the segment intercepted on either of the parallel tangents, between the chord of contact of the other two and the curve, is bisected by the non-parallel tangent.

Proposition XXIV.

If a circle intersect an ellipse in four points, the common chords will be equally inclined to the axis.

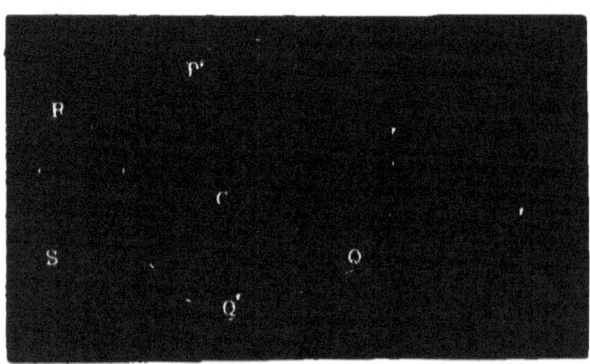

Fig. 34.

Let PP', QQ' be the common chords intersecting in O.

Draw the semi-diameters CR, $CS \parallel PP'$, QQ' respectively.

Then $OP \times OP' : OQ \times OQ' = CR^2 : CS^2$;

(Prop. xxii.)

but $OP \times OP' = OQ \times OQ'$; (35, III Euclid.)

$\therefore CR = CS.$

Hence CR and CS, and $\therefore OP'$ and OQ', are equally inclined to the axis. (*Cor.* 1, Prop. i.)

Cor. 1.—In like manner it can be shown that PQ and $P'Q'$, also PQ' and $P'Q$ are equally inclined to the axis.

Proposition XXV.

If a tangent to an ellipse at any point P meet any diameter AA' produced in T, and the ordinate PM be drawn; then CA is a mean proportional between CM and CT.

Fig. 35.

Draw tangents at A and A', meeting the tangent at P in G and H.

Then $\qquad A'T : TA = A'H : AG$; (Parallel lines.)
and $\qquad A'M : MA = HP : PG$;
but $\qquad A'H : AG = HP : PG$;
$\qquad\qquad\qquad\qquad\qquad$ (Cor. 3, Prop. xxii.)
$\qquad \therefore A'T : TA = A'M : MA$;
hence $\qquad CM \times CT = CA^2$.

Cor. 1.—Conversely, if PM be an ordinate to any diameter AA', and CT be taken a third proportional to CM and CA; then PT will be a tangent to the ellipse.

Cor. 2.—The tangents at the extremities of any double ordinate intersect on the diameter corresponding to that ordinate.

Cor. 3.—If a diameter be drawn through the intersection of two tangents to an ellipse, it will bisect the chord of contact.

Proposition XXVI.

If a variable tangent to an ellipse meet two fixed parallel tangents, it will intercept segments on them whose rectangle is constant, and equal to the square of the parallel semi-diameter.

(See Fig. 35.)

Let GPH be the variable tangent meeting the fixed tangents AG, $A'H$ in the points G, H.

Draw PE an ordinate to BC.

Then $\qquad CT : CA = CA : CM.$ $\qquad$ (Prop. xxv.)

Division $\qquad CT : TA = CA : AM.$

Alt. $\qquad CT : CA = TA : AM.$

Comp. $\qquad CT : TA' = TA : TM;$

$\qquad \therefore CD : A'H = AG : PM;$ $\qquad$ (Similar $\triangle^s$.)

$\qquad \therefore A'H \times AG = CD \times PM$

$\qquad\qquad\qquad\quad = CD \times CE = CB^2.$ (Prop. xxv.)

Cor. 1.—The rectangle under the segments of the variable tangent is equal to the square of the semi-diameter CQ drawn parallel to it.

For $\qquad A'H : HP = CB : CQ = AG : PG;$
$\qquad\qquad\qquad\qquad\qquad\qquad$ (*Cor.* 3, Prop. xxii.)

$\qquad \therefore A'H \times AG : HP \times PG = CB^2 : CQ^2.$

But $\qquad\qquad A'H \times AG = CB^2;$

$\qquad \therefore HP \times PG = CQ^2.$

CHAP. II.]   *The Ellipse.*   87.

Cor. 2.—Any tangent to an ellipse will be cut harmonically by two parallel tangents and the diameter passing through their points of contact.

For $\quad HT : TG = A'H : AG \quad$ (Similar $\triangle$'s.)

$$= HP : PG.$$
$$(Cor.\ 3,\ \text{Prop. xxii.})$$

Cor. 3.—If two parallel tangents AC, BF, be cut by any two other tangents EF, CD; then

$$AE : BD = EO : OF = CO : OD.$$

Fig. 36.

For $\quad CA \times BD = AE \times BF; \quad$ (Prop. xxvi.)

$$\therefore AE : AC = BD : BF;$$

$$\therefore AE : AC - AE = BD : BF - BD.$$

Alt. $\quad AE : BD = EC : DF$

$$= EO : OF,\ \text{or} = CO : OD.$$

Cor. 4.—The lines joining F with C, and D with E, would, if produced, intersect on the diameter BA produced.

Proposition XXVII.

The triangles CPT and CAK, formed by drawing tangents at the extremities of any two semi-diameters of an ellipse, are equal in area.

Fig. 37.

Draw the ordinate PM.
Then $\qquad TC : CA = CA : CM.$ (Prop. xxv.)
$\qquad\qquad\qquad = CK : CP;$ (Parallel lines.)
$\therefore KT$ is $\parallel PA$;
$\therefore \triangle PTA = \triangle PKA$; (37, I. Euclid.)
$\therefore \triangle CPT = \triangle CAK.$

Cor. 1.—If the ordinate AN be drawn from A to the semi-diameter CP, then the area of the $\triangle CMP$ = area of $\triangle CAN$.

For $\qquad AN$ is $\parallel PT$;
$\therefore TC : CA = PC : CN$;
hence $\qquad CA : CM = PC : CN$;
$\therefore MN$ is $\parallel PA$;
$\therefore \triangle NPM = \triangle NAM$; (37, I. Euclid.)
$\triangle CMP = \triangle CAN.$

Proposition XXVIII.

If from the extremities of any two conjugate diameters CP, CQ, the ordinates PM, QN be drawn to any other diameter CA; then

$$CN^2 = AM \times MA', \text{ and } CM^2 = AN \times NA'.$$

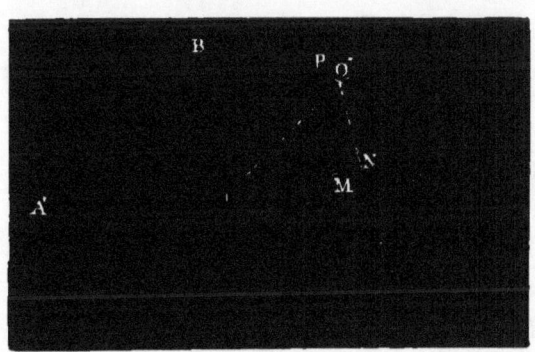

Fig. 38.

$$MT^2 : CN^2 = MP^2 : QN^2 \qquad \text{(Similar } \triangle^s.)$$
$$= MA \times MA' : NA \times NA'.$$
$$\text{(Cor. 2, Prop. xviii.)}$$

But $\quad MA \times MA' = AC^2 - CM^2 \qquad$ (5, II. Euclid.)
$$= TC \times CM - CM^2 \qquad \text{(Prop. xxv.)}$$
$$= CM \times MT;$$

$\therefore MT^2 : CN^2 = CM \times MT : AN \times NA'.$

Alt. $\quad MT^2 : CM \times MT = CN^2 : CA^2 - CN^2;$
$$\text{(5, II. Euclid.)}$$

$\therefore MT : CM = CN^2 : CA^2 - CN^2.$

Comp. $\quad MT : CT = CN^2 : CA^2;$

$\therefore CM \times MT : CM \times CT = CN^2 : CA^2.$

But $CM \times CT = CA^2$; (Prop. xxv.)

$\therefore CN^2 = CM \times MT = AM \times MA'$.

In like manner it can be proved that $CM^2 = AN \times NA'$.

Cor. 1.— $CM^2 + CN^2 = CA^2$.

For
$$CM^2 + CN^2 = AN \times NA' + AM \times MA'$$
$$= CA^2 - CN^2 + CA^2 - CM^2;$$
(5, II. Euclid.)

$\therefore 2CM^2 + 2CN^2 = 2CA^2$;

or, $CM^2 + CN^2 = CA^2$.

Cor. 2.—$PM^2 + QN^2 = BC^2$, BC being the semi-diameter conjugate to CA.

For $CB^2 : CA^2 = PM^2 : AM \times MA'$ or CN^2;
(Prop. xviii.)

also, $CB^2 : CA^2 = QN^2 : AN \times NA'$ or CM^2;

$\therefore CB^2 : CA^2 = PM^2 + QN^2 : CN^2 + CM^2$.

But $CA^2 = CN^2 + CM^2$, $\therefore CB^2 = PM^2 + QN^2$.

Cor. 3.—$CA : CB = CM : QN = CN : PM$.

For $CA^2 : CB^2 = AM \times MA'$, or $CN^2 : PM^2$
$$= AN \times NA', \text{ or } CM^2 : QN^2;$$

$\therefore CA : CB = CN : PM = CM : QN$.

Cor. 4.—The $\triangle CMP$ is $= \triangle CNQ$ in area.

Produce QN to Q'. Then $QN = Q'N$;

$\therefore CN : CM = PM : Q'N$; (*Cor.* 3.)

$\therefore \triangle CMP = \triangle CNQ' = \triangle CNQ$.
(15, VI. Euclid.)

Proposition XXIX.

If any tangent to an ellipse meet any two conjugate diameters CP, CQ, the rectangle under its segments is equal to the square of the parallel semi-diameter.

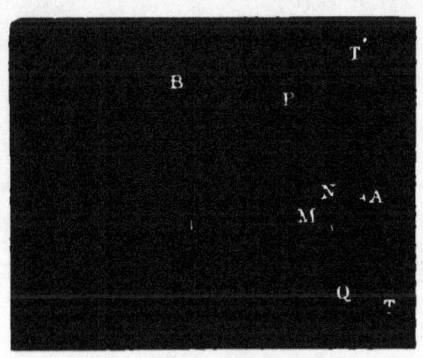

Fig. 39.

Draw the ordinates, PM, QN, to the diameter passing through the point of contact of the tangent.

Then $\quad CM : PM = CA : AT'$; (Similar $\triangle^s$.)

$\therefore CM \times AT' = PM \times CA$

$\qquad = CB \times CN;$
(*Cor.* 3, Prop. XXVIII.)

$\therefore CM : CN = CB : AT'$.

Again, $\quad CN : NQ = CA : AT$; (Similar $\triangle^s$.)

$\therefore CN \times AT = CA \times NQ$

$\qquad = CB \times CM;$
(*Cor.* 3, Prop. XXVIII.)

$\therefore CM : CN = AT : CB$;

hence $CB : AT' = AT : CB$;

$\therefore AT \times AT' = CB^2$.

Otherwise thus:—

Fig. 40.

Draw PM, QM' ordinates to CA, and PN an ordinate to CB.

Then $CS \times CM' = CA^2 = CM \times CT$; (Prop. xxv.)

$\therefore CS : CT = CM : CM'$.

But $CS : CT = CQ : PT$; (Similar Δ^s.)

$\therefore CQ : PT = CM : CM'$

$= PN : CM'$

$= PT' : CQ$; (Similar Δ^s.)

$\therefore PT \times PT' = CQ^2$.

Cor.—Conversely, if on any tangent segments AT, AT' be measured from the point of contact whose rectangle is equal to the square of the parallel semi-diameter, then the diameters drawn through T and T' will be conjugate.

Proposition XXX.

Given in magnitude and position any two conjugate semi-diameters CA', CB', of an ellipse, to find the axes.

Fig. 41.

On CB' produced take BD a third proportional to CB' and CA'; bisect CD in H; draw $HO \perp CD$ to meet $B'O$ drawn $\parallel CA'$; with the centre O and radius OC describe a circle cutting $B'O$ produced in T and T'; join CT, CT'. Draw $B'M$ and $B'N \parallel CT'$ and CT respectively. Take CA a mean proportional between CT and CM; also CB a mean proportional between CT' and CN. Then CA and CB are the semi-axes.

For $\quad CA'^2 = CB' \times B'D \quad$ (Const.)

$\quad\quad\quad = TB' \times B'T'; \quad$ (35, III. Euclid.)

∴ CA and CB are conjugate diameters;
(Cor., Prop. xxix.)

but $\angle TCT' = 90°$; hence CA and CB are the semi-axes.

94   The Ellipse.   [CHAP. II.

Proposition XXXI.

If a straight line cut two tangents TP, TQ to an ellipse in the points A, A', the curve in B, B', and the chord of contact in O; then

$$AO^2 : A'O^2 = AB \times AB' : A'B \times A'B'.$$

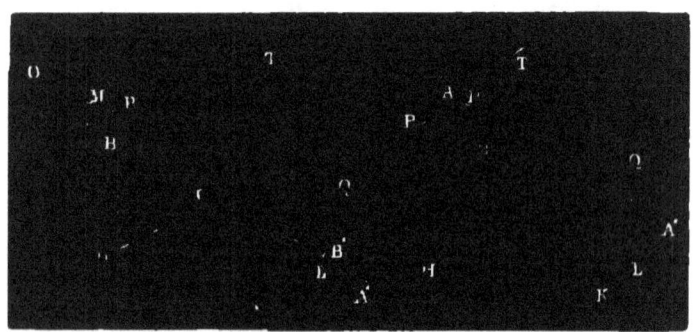

Fig. 42.   Fig. 43.

Draw $AM \parallel TQ$; also the semi-diameters CH, CK, and $CL \parallel TP$, TQ and AA', respectively.

Then $\quad AO^2 : A'O^2 = AM^2 : A'Q^2 \quad$ (Parallel lines.)

$$= \begin{cases} AM^2 : AP^2 \\ AP^2 : A'Q^2 \end{cases}$$

$$= \begin{cases} TQ^2 : TP^2 \\ AP^2 : A'Q^2 \end{cases} \quad \text{(Similar } \Delta^s.)$$

$$= \begin{cases} CK^2 : CH^2 \\ AP^2 : A'Q^2 \end{cases}$$
(Cor. 2, Prop. XXII.)

$$= \begin{cases} CK^2 : CL^2 \\ CL^2 : CH^2 \\ AP^2 : A'Q^2 \end{cases}$$

[CHAP. II.] *The Ellipse.* 95

$$= \left\{ \begin{array}{c} A'Q^2 : A'B \times A'B' \\ AB \times AB' : AP^2 \\ AP^2 : A'Q^2; \end{array} \right\}$$

(*Cor.* 4, Prop. XXII.)

$$\therefore AO^2 : A'O^2 = AB \times AB' : A'B \times A'B'.$$

Proposition XXXII.

If any line TS be drawn parallel to the chord of contact of the tangents to an ellipse, the segments AT, BS intercepted between the curve and the tangents will be equal.

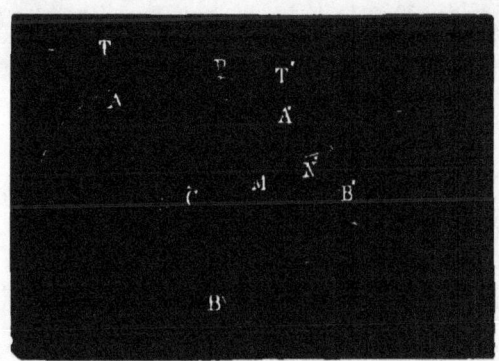

Fig. 44.

Draw the diameter bisecting the parallel chords PQ, AB. This diameter will pass through O.

(*Cor.* 2, Prop. XXV.)

Then, since $\qquad PM = MQ$;

$\qquad\qquad \therefore TN = NS$;

but $\qquad\qquad AN = NB$;

$\qquad\qquad \therefore AT = BS.$

Proposition XXXIII.

Any line OB, drawn through the intersection of two tangents to an ellipse, is cut harmonically by the curve and the chord of contact of the tangents.

Fig. 45.

Through B and B' draw TS and $T'S'$ ∥ PQ, the chord of contact of the tangents.

Then $TP^2 : T'P^2 = TB \times TA : T'B' \times T'A'$;
(*Cor.* 4, Prop. XXII.)

but $TB = AS$, and $T'B' = A'S'$; (Prop. XXXII.)

∴ $TP^2 : T'P^2 = TB \times BS : T'B' \times B'S'$

$= OB^2 : OB'^2$; (Similar △^{s}.)

but $TP : T'P = O'B : O'B'$; (Parallel lines.)

hence $OB : OB' = O'B : O'B'$.

Cor.—If through any point O without an ellipse a line be drawn cutting the curve in B, B', and OO' be taken equal to the harmonic mean between OB and OB', the locus of O' is a straight line; namely, the chord of contact of the tangents drawn from O.

CHAP. II.] *The Ellipse.* 97

Proposition XXXIV.

Any line drawn through the middle point of the chord of contact of two tangents to an ellipse, will be cut harmonically by the curve, and the line drawn through the intersection of the tangents parallel to their chord of contact.

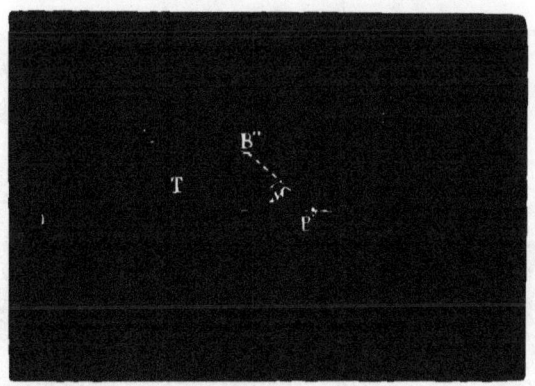

Fig. 46.

Let TP, TQ be tangents, BB' any chord drawn through O', the middle point of PQ, and meeting the line drawn through $T \parallel PQ$ in O. Through B'' draw $B'B'' \parallel PQ$; join TC.

Then TC will pass through O'. (*Cor.* 3, Prop. xxv.)

$$B''M : B''B' = \begin{Bmatrix} B''M : O'T'' \\ O'T' : B''B' \end{Bmatrix}$$

$$= \begin{Bmatrix} B''T : TT' \\ BT' : BB'' \end{Bmatrix} \quad \text{(Similar } \triangle\text{'s.)}$$

$$= B''T \times BT' : TT' \times BB''.$$

But $TT' \times BB'' = 2B''T \times BT'$;

∴ $B'B'' = 2B''M$; ∴ $B'M$ is an ordinate, and B' a point on the curve.

Now $BT : TB'' = BT' : T'B''$; (Prop. xxxiii.)

∴ $BO : OB' = BO' : O'B'$. (Parallel lines.)

H

Otherwise thus:—

Let TP, TQ be tangents, BB' any chord drawn through O', the middle point of PQ, and produced to meet the line drawn through $T \parallel PQ$ in O; join OC, TC; then (*Cor.* 3, Prop. xxv.) TC will pass through O'. Draw RN a tangent at R, $RM \parallel QP$, and $\therefore$ an ordinate to CT; through O' draw $DE \parallel RN$.

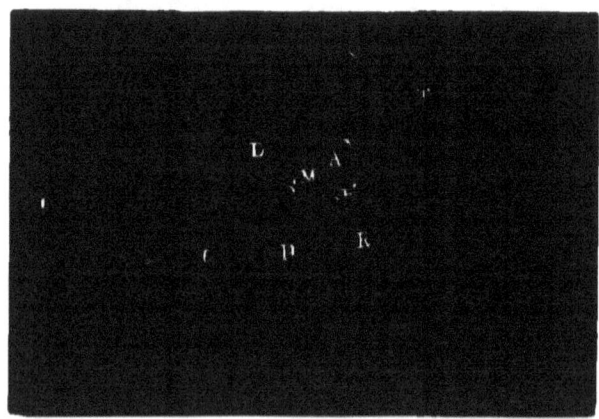

Fig. 47.

Then, $CT \times CO' = CA^2 = CN \times CM$; (Prop. xxv.)

$\therefore CT : CM = CN : CO'$;

$\therefore CO : CR = CR : CH$; (Parallel lines.)

$\therefore$ the tangents at D and E will intersect at O;

(Prop. xxv.)

and hence $BO : OB' = BO' : O'B'$.

(Prop. xxxiii.)

Cor. 1.—If through any point O' within an ellipse any line be drawn cutting the curve in B, B', and O be taken the harmonic conjugate to O' with regard to B and B', the locus of O is a straight line.

Cor. 2.—If from O tangents be drawn to the ellipse, the chord of contact will pass through O'.

For (Prop. xxxiii.) the chord of contact must divide the line OB harmonically.

[CHAP. II.] *The Ellipse.* 99

Proposition XXXV.

If through any point P on an ellipse lines POO', $Q'PQ$ be drawn parallel to any two adjacent sides DA, DC of an inscribed quadrilateral, meeting the opposite sides in O, O', and Q, Q'; then $PO \times PO' : PQ \times PQ'$ in a constant ratio.

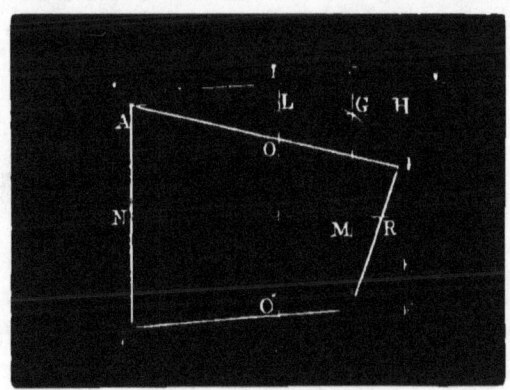

Fig. 48.

Through B and C draw BE and $CG \parallel AD$; join AG, and produce it to meet EB produced in H; draw the diameter NMR bisecting the parallel chords AD, GC, BE.

Then $HR = RK$;   (Parallel lines.)

hence $BH = EK$

Now $OL : HB = LA : AH$

$\qquad\qquad\quad = O'D : DK$.

Alt. $OL : O'D = HB : DK$.

Also PO or $SC : QS = BK : KC$.   (Similar $\triangle^s$.)

Compounding,

$OL \times PO' : O'D \times QS = HB \times BK : DK \times KC$;

H 2

100   *The Ellipse.*   [CHAP. II.

$$\therefore OL \times PO' : PQ' \times QS = EK \times BK : DK \times KC$$
$$= O'V \times O'P : O'C \times O'D$$
$$= PL \times O'P : PS \times PQ';$$
$$\therefore OL \times PO' + PL \times PO' : PQ' \times QS + PS \times PQ'$$
$$= EK \times BK : DK \times KC,$$
or $PO \times PO' : PQ \times PQ' = EK \times BK : DK \times KC.$

Now it is evident that the points A, B, C, D, being fixed, E is also fixed, and

$\therefore EK \times BK : DK \times KC$ in a constant ratio;

$\therefore PO \times PO' : PQ \times PQ'$ in a constant ratio.

Proposition XXXVI.

If from any point P on an ellipse, lines PR, PR', PS, PS', be drawn to the sides of an inscribed quadrilateral, making with them any constant angles, then the rectangles under the lines drawn to the opposite sides will be in a constant ratio.

Fig. 49.

Take any other point p on the curve.

CHAP. II.]   *The Ellipse.*   101

Through the points P and p draw QPQ' and qpq' ∥ DC, and POO', poo' ∥ AD, also pr, pr', ps, ps' ∥ PR, PR', PS, PS', respectively.

Then $\quad\quad\quad PR : pr = PQ : pq,\quad$ (Similar △ˢ.)

and $\quad\quad\quad PR' : pr' = PQ' : pq'$; (Do.)

∴ $PR \times PR' : pr \times pr' = PQ \times PQ' : pq \times pq'$.

Similarly it may be proved that

$$PS \times PS' : ps \times ps' = PO \times PO' : po \times po';$$

but $\quad PQ \times PQ' : PO \times PO' = pq \times pq' : po \times po'$;
$\quad\quad\quad\quad\quad\quad\quad\quad\quad\quad\quad$ (Prop. xxxv.)
∴ $\quad PR \times PR' : PS \times PS' = pr \times pr' : ps \times ps'$.

Cor. 1.—The rectangle under the perpendiculars let fall from any point of an ellipse on two opposite sides of an inscribed quadrilateral is in a constant ratio to the rectangle under the perpendiculars let fall on the other two sides.

Cor. 2.—If the points A and B coincide, also the points C and D, then the sides AB and CD become tangents, and the sides BC and AD coincide, and become the chord of contact. Then the rectangle under the perpendiculars let fall from any point of an ellipse on two fixed tangents is in a constant ratio to the square of the perpendicular let fall on their chord of contact.

Cor. 3.—If we suppose AB, CD to intersect in X and AC, BD in Y; also that PR, PR', PS, PS', pr, pr', ps, ps' lie in the line XY; then, regarding $ACBD$ as the quadrilateral, the above proportion becomes

$$PX^2 : PY^2 = pX^2 : pY^2 ;$$

∴ XY is cut harmonically by the curve.

Proposition XXXVII.

If two fixed tangents AD, DB, to an ellipse be cut by a diameter AB, parallel to their chord of contact, and by a third variable tangent EF, the rectangle under the segments of the two fixed tangents, intercepted between the diameter and the variable tangent, is constant.

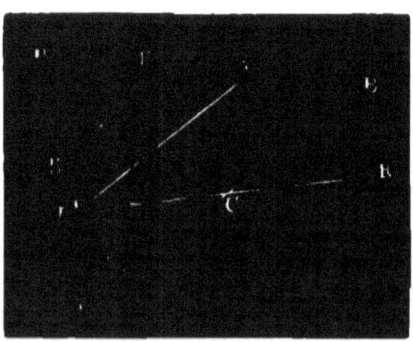

Fig. 50.

Join PC; produce it to meet the curve in R; join BR.

Since $\qquad CA = CB$; $\qquad$ (Prop. xxxii.)

∴ the $\triangle^s ACP$ and BCR are equal; (4, I. Euclid.)

∴ BR is = and ∥ AP;

∴ BR is a tangent.

Then $\quad RB : PE = BF : FD$; $\quad$ (Cor. 2, Prop. xxvi.)

∴ $AP : PE = BF : FD$;

∴ $AP : AP + PE = BF : BF + FD$;

∴ $AP : AE = BF : BD$;

∴ $AE \times BF = BD \times AP$, and ∴ constant.

Lemma.

If on any right line we have segments AB, BC, &c., and on any other right line inclined to the former at any angle segments $A'B'$, $B'C'$, &c., such that $AB : A'B' = AC : B'C'$, &c. $= m : n$; then the locus of the middle points of AA', BB', CC', &c., will be a right line.

Fig. 51.

Measure off $C'O'$ and AO'', such that

$$CO : C'O' = AO'' : A'O = m : n.$$

Bisect OO'' in M and OO' in M'; join MM'.
This line will pass through the middle points of AA', BB', and CC'.

For $AB : A'B' = BC : B'C' = CO : C'O' = AO : A'O$;

$\therefore \frac{1}{2}(AB + BC + CO + AO'') : \frac{1}{2}(A'B' + B'C' + C'O' + A'O)$

$$= m : n;$$

or $\qquad OM : OM' = m : n.$

Again, $AB + AO'' : A'B' + A'O = m : n = MO'' : M'O$;

$\therefore AB + AO'' - MO'' : A'B' + A'O - M'O = m : n$;

or $\qquad MB : M'B' = m : n.$

Now consider the △ OBB', the sides of which are cut by the transversal MM';

$$\therefore \frac{BF}{FB'} \cdot \frac{B'M'}{M'O} \cdot \frac{OM}{MB} = 1,$$

or $$\frac{BF}{FB'} \cdot \frac{B'M'}{MB} \cdot \frac{OM}{M'O} = 1;$$

$$\therefore \frac{BF}{FB'} \cdot \frac{n}{m} \cdot \frac{m}{n} = 1;$$

$$\therefore BF = FB'.$$

In like manner it can be shown that AA' and CC' are bisected by MM'.

Proposition XXXVIII.

The right line joining the middle points of the diagonals of a quadrilateral circumscribing an ellipse will pass through the centre.

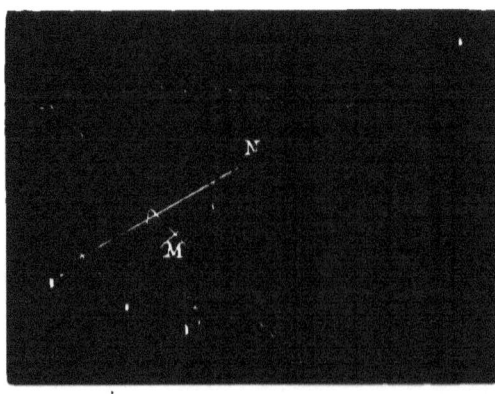

Fig. 52.

Let $ABA'B'$ be the quadrilateral. Through the centre C draw $DD' \parallel PQ$, the chord of contact of the opposite sides AB, $A'B'$.

CHAP. II.   *The Ellipse.*   105

Then $AD \times B'D' = BD \times A'D'$;
                                    (Prop. xxxvii.)
∴ $AD : A'D' :: BD : B'D'$.

Hence M, N, and C, the middle points of AA', BB', and DD', lie on a right line.
                    (Prop. xxxii., and Lemma, page 103.)

Proposition XXXIX.

If a quadrilateral be circumscribed to an ellipse, the diagonals will intersect on the chords of contact of the sides.

Fig. 53.

For AC will be cut by PQ, the chord of contact of the tangents DA, BC, so that

$$AO^2 : OC^2 = AH \times AG : CG \times CH ;$$
                                    (Prop. xxxi.)
but it will also be cut by the chord of contact of the tangents AB, CD in the same ratio;

∴ PQ and RS will intersect AC in the same point.

In like manner it may be shown that the diagonal DB will pass through the intersection of PQ and RS.

106   *The Ellipse.*   [CHAP. II.

PROPOSITION XL.

The areas of the ellipse and auxiliary circle are in the ratio $CB : CA$.

Let the ellipse and auxiliary circle be divided by any number of lines drawn parallel to the axis minor.

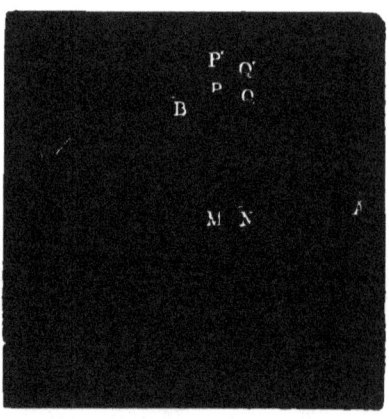

Fig. 54.

Then, since

$$PM : P'M = QN : Q'N = CB : CA;$$

(*Cor.* 3, Prop. VIII.)

∴ quadrl. $PMNQ$: quadrl. $P'MNQ' = CB : CA$.

The same will be true of all the figures similarly described in the ellipse and auxiliary circle.

Hence the sum of all the quadrilateral figures inscribed in the ellipse : the sum of all the corresponding quadrilateral figures inscribed in the circle $= BC : CA$.

Now this will be true whatever be the number of the figures.

Let the number of the figures be increased, and the breadth of each indefinitely diminished. Then the sum of the areas of the figures inscribed in the ellipse will be

CHAP. II.]   *The Ellipse.*   107

equal to the area of the ellipse, and the sum of the areas of those inscribed in the circle equal to the area of the circle;

∴ area of ellipse : area of circle = $CB : CA$.

Proposition XLI.

The section of a right cone by a plane, which intersects all the generating lines on the same side of the vertex, will be an ellipse.

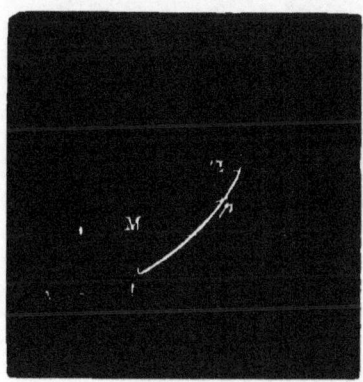

Fig. 55.

Let the plane BVE, drawn through the axis of the cone perpendicular to the plane of the *section*, coincide with the plane of the paper; then both the *section* APA' and the base BPE will be perpendicular to the plane of the paper; therefore the line MP, in which the section cuts the base, is perpendicular to the plane of the paper, and therefore perpendicular to BE, the diameter of the base; hence

$$BM \times ME = MP^2. \quad \text{(35, III. Euclid.)}$$

If, now, any other plane bpe be drawn parallel to the base, meeting the section in pm, it can similarly be shown that mp is perpendicular to be; ∴ $bm \times me = mp^2$.

108   The Ellipse.   [CHAP. II.

But $BM : bm = A'M : A'm$; (Similar $\triangle^s$.)
also $ME : me = MA : mA$;
∴ $BM \times ME : bm \times me = A'M \times MA : A'm \times mA$,
or $MP^2 : mp^2 = A'M \times MA : A'm \times mA$.

Hence the section APA' is an ellipse. (Prop. IX.)

Fig. 56.

Cor. 1.—The axis minor is a mean proportional between the diameters of the frustrum of the cone which just includes the section.

Draw the section EBG parallel to the base, bisecting AA' in C.

Then $CE = \frac{1}{2} AD$ and $CG = \frac{1}{2} A'D'$; (Similar $\triangle^s$.)
∴ $CE \times CG$ or $CB^2 = \frac{1}{4} AD \times A'D'$;

hence $4CB^2$ or the square of the axis minor $= AD \times A'D'$.

Cor. 2.—The Latus-rectum is a fourth proportional to AA', AD and $A'D'$.

Since ABA' is an ellipse,

$$L \times AA' = 4BC^2 \quad (Cor. 1, Prop. \text{IX.})$$
$$= AD \times A'D' ;$$
∴ $AA' : AD = A'D' : L$.

CHAP. II.] *The Ellipse.* 109

Cor. 3.—If CF and CF' be measured off $= AG$ or $\frac{1}{2}AD'$, then F and F' are the foci.

Draw $\quad AH \perp A'D'$.

Then $\quad AA'^2 - AD'^2 = A'H^2 - HD'^2$

$$= A'D' \times AD$$
$$\qquad\qquad (Cor., 5, \text{II. Euclid.})$$
$$= 4BC^2; \qquad (Cor.\ 1.)$$

$\therefore \quad CA^2 - AG^2 = BC^2;$

hence $\quad CA^2 - BC^2 = AG^2 = CF^2 = CF'^2;$

$\therefore F$ and F' are the foci. $\qquad (Cor.\ 3,\ \text{Prop. II.})$

Cor. 4.—The spheres inscribed in a cone, to touch the plane of the section, will determine the foci.

For $\quad AF = AC - CF$

$$= AC - AG \qquad\qquad (Cor.\ 3.)$$
$$= \tfrac{1}{2}(AA' - A'D)$$
$$= \tfrac{1}{2}(AA' + AV - VD - A'D)$$
$$= \tfrac{1}{2}(AA' + AV - VA');$$

$\therefore F$ is the point of contact of the circle inscribed in the triangle $AA'V$. (See Galbraith and Haughton's Manual of Euclid, Appendix, Book IV.)

In like manner it may be shown that F' is the point of contact of the circle exscribed to the triangle $AA'V$.

Problems on the Ellipse.

1. The axis major is the longest straight line that can be drawn in an Ellipse.

Join the two foci with the extremities of the chord; then twice the chord will be less than the sum of the four focal radii vectores, and therefore less than twice the axis major.

2. The minor axis is the least diameter of an Ellipse.

Connect Problem 1 with *Cor.* 5, Prop. xvi.

3. Given the foci of an Ellipse, describe it so as to touch a given line.

See Prop. iv.

4. Find the locus of the point of intersection of any tangent to an Ellipse, with the line drawn from the focus making a constant angle with the tangent.

If the vertex of a triangle of a given species be fixed, while one base angle moves along a fixed circle, the locus of the other base angle will be a circle. See also Prop. vi.

5. The angle between the tangents at the extremities of a chord which passes through either focus is half the supplement of the angle which the chord subtends at the other focus.

Draw the normals at the extremities of the focal chord, and see *Cor.* 3, Prop. iii.

6. Diameters parallel to supplemental chords are conjugate.

See Def., page 71, and *Cor.* 2, Prop. xvii.

7. If a rectangle be circumscribed to an Ellipse, its diagonals determine the equi-conjugate diameters.

8. Show that Prop. xxxiii. follows from Prop. xxxvi.

9. The line drawn parallel to the axis major through the intersection of normals at the extremities of a focal chord will bisect the chord.

See Prop. xiii.; also note properties of circles inscribed, exscribed, and circumscribed to a triangle.

10. When is the sum of two conjugate diameters a minimum; and when is it a maximum?

See *Cor.* 5, Prop. xvi.; *Cor.*, Prop. xxi., and Prop. xxix.

CHAPTER III.

THE HYPERBOLA.

DEFINITIONS.

A HYPERBOLA is the curve traced out by a point which moves in such a way that its distance from a fixed point is to its distance from a fixed right line in a constant ratio $\epsilon : 1$ (ϵ being greater than unity).

The fixed point is called the *Focus*, and the fixed right line the *Directrix*.

For Definitions of *Axis, Vertex, Centre, Diameter, Tangent, Normal, Chord, Ordinate, Abscissa, Latus rectum*, &c., see Definitions, Chapters I. and II.

Conjugate Diameters may be defined, as on page 72. *Supplemental Chords, Transverse Axis, Conjugate Axis*, as on page 71.

Two hyperbolas are said to be *conjugate* when the transverse axis of each is the conjugate axis of the other.

If the axes be equal the hyperbola is said to be *Rectangular* or *Equilateral*.

Tangents to the curve whose points of contact are at infinity are called *Asymptotes*.

Proposition I.

The focus, directrix, and eccentricity of a hyperbola being given, to determine any number of points on the curve.

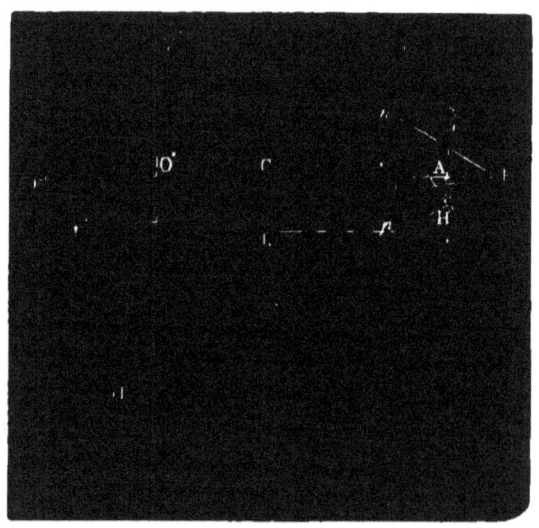

Fig. 1.

Let F be the focus, and Oy the directrix.

Draw $FO \perp$ the directrix. Divide FO internally at A, and externally at A', so that $FA : AO = FA' : A'O$ in the given ratio $\epsilon : 1$; then A and A' are the vertices of the curve, and AA' the axis.

On the directrix take any point p; join FP, and produce it; draw AH, $A'H'$ perpendiculars to AA', meeting Fp in H, H'. On HH' as diameter describe a circle; through p draw PpP' ∥ to the axis, cutting the circle in P, P'; join PF, $P'F$.

Then, since AH is ∥ Op; ∴ $HF : Hp = AF : AO$.

Also, since $A'H'$ is ∥ Op; ∴ $H'F : H'p = A'F : A'O$;

∴ $H'F : H'p = HF : Hp$.

CHAP. III.]   *The Hyperbola.*   113

Hence $PF : Pp = FH : Hp$ (Lemma, page 47.)

$\qquad\qquad = FA : AO = \epsilon : 1.$ (2, VI. Euclid.)

Also $P'F : P'p = FH : Hp$ (Lemma, page 47.)

$\qquad\qquad = FA : AO = \epsilon : 1.$ (2, VI. Euclid.)

∴ P and P' are points on the curve.

In like manner, by taking other points on the directrix, any number of points on the curve may be determined.

Cor. 1.—If Op' be taken equal to Op, and another point Q found in a similar manner to P, it is obvious that Qp' will be equal to Pp. Therefore, corresponding to any point P on the curve, at a perpendicular distance from the axis = Op, there is another point Q on the other side of the axis, at a distance from it = Op, and also at the same distance as P from the directrix; hence the curve is symmetrical with regard to the axis.

Cor. 2.—Bisect AA' in C; through C draw $CLS \perp AA'$, and ∴ $\perp PP'$, meeting HH' in S, and PP' in L.

Then, since AH, CS and AH' are parallel, ∴ $SH = SH'$;

∴ S is the centre of the circle $H'P'HP$;

∴ $LP = LP'.$ (3, III. Euclid.)

Therefore, corresponding to any point P on the curve, there is another point P' on the other side of CL, situated in precisely the same manner with regard to it as P; hence the curve is symmetrical also with regard to CL.

If therefore CO' be measured off = CO, and $CF' = CF$, and $O'Y'$ be drawn $\perp O'C$, the curve could be equally well described with F' as focus, and $O'Y'$ as directrix.

Cor. 3.—Since $PF : Pp = AF : AO = A'F : A'O$;

∴ $PF : Pp = A'F - AF : A'O - AO$

$\qquad\qquad = 2CA : 2CO;$

∴ $PF : Pp = CA : CO.$

I

114 *The Hyperbola.* [CHAP. III.

That is, the distance of any point on the curve from the focus is to its distance from the directrix as $CA : CO$.

Cor. 4.—Since $A'F : A'O = AF : AO$;

$$\therefore A'F + AF : A'F - AF = A'O + AO : A'O - AO;$$

or $$2CF : 2CA = 2CA : 2CO;$$

$$\therefore CF \times CO = CA^2.$$

Cor. 5.—From the symmetry of the curve it is evident that any line drawn through C to meet the curve will be bisected at that point; hence C is the centre of the curve.

PROPOSITION II.

The difference of the distances of the foci from any point on a hyperbola is equal to the transverse axis.

Fig. 2.

$$PF' : Pp' = CA : CO; \quad (Cor.\ 3, \text{Prop. 1.})$$

also $$PF : Pp = CA : CO;$$

$$\therefore PF' - PF : Pp' - Pp = CA : CO;$$

but $$Pp' - Pp = OO' = 2CO;$$

$$\therefore PF' - PF = 2CA = AA'.$$

CHAP. III.]   *The Hyperbola.*   115

Cor. 1.—By the help of the preceding theorem we can construct a hyperbola mechanically.

If two of the extremities of two threads, the difference of whose lengths is equal to AA', be fastened to the two fixed points A, A', and the other extremities, having been passed through the eye of a needle, be joined, it is obvious that the eye of the needle, moved about so as to keep the threads always stretched, will describe a hyperbola whose foci are F and F'.

Cor. 2.—The difference of the distances of any point from the focii of a hyperbola is greater or less than the transverse axis, according as the point is on the concave or convex side of the curve.

Fig. 3.

First let Q be any point on the concave side of the curve.

Join QF, QF', and let QF' meet the curve in P; join PF.

Then $\qquad QF' = QP + PF'$,

and $\qquad QF < QP + PF$; $\qquad$ (20, I. Euclid.)

$\qquad \therefore QF' - QF > PF' - PF$;

but $\qquad PF' - PF = AA'$; $\therefore QF' - QF > AA'$.

116 *The Hyperbola.* [CHAP. III.

Next, let Q' be on the convex side.

Join $Q'F$, $Q'F'$; produce $Q'F'$ to meet the curve in P; join PF.

Then $\qquad F'Q' = F'P - PQ'$,

and $\qquad Q'F > PF - PQ'$; (*Cor.*, 20, I. Euclid.)

$\qquad \therefore Q'F' - Q'F < PF' - PF$;

but $\qquad PF' - PF = AA'$; $\therefore Q'F' - Q'F < AA'$.

Cor. 3.—Conversely, a point will be on the concave or convex side of a hyperbola, according as the difference of its distances from the foci is greater or less than the transverse axis.

Proposition III.

1°. The line which bisects the angle formed by drawing lines from any point P on a hyperbola to the foci lies altogether without the curve.

2°. Any other line drawn through the point P will cut the curve.

Fig. 4.

1°. Let PH bisect the ∠ between PF and PF'.

Cut off $PR = PF$; join F, F' and R with *any* point Q on the line PH.

CHAP. III.]   *The Hyperbola.*   117

Then the △^{s} QFR and QRP are equal;   (4, I. Euclid.)
$$\therefore QR = QF;$$
but $F'Q - QR < F'R$;   (*Cor.*, 20, I. Euclid.)
$$\therefore F'Q - QF < F'R$$
$$< F'P - PR$$
$$< F'P - PF \qquad \text{(Const.)}$$
$$< AA'; \qquad \text{(Prop. II.)}$$
$\therefore$ Q is a point without the curve.   (*Cor.* 3, Prop. II.)

2°. Let PK be any line other than the bisector of the $\angle FPF'$.

Fig. 5.

Draw PL, making the $\angle KPL = \angle KPF$; cut off $PL = PF$.

Join $F'L$; produce $F'L$ to meet PK in Q; join QF.

Then the △^{s} FPQ and LPQ are equal;   (4, I. Euclid.)
$$\therefore QF = QL;$$
but $F'L > F'P - PL$
$$\text{(}Cor.\text{, 20, I. Euclid.)}$$
$$> F'P - PF$$
$$> AA'; \qquad \text{(Prop. II.)}$$
$$\therefore F'Q - QL \text{ or } F'Q - QF > AA';$$
$\therefore$ Q is a point on the concave side of the curve.
(*Cor.* 3, Prop. II.)

118   *The Hyperbola.*   [CHAP. III.

Cor. 1.—Hence the line which bisects the angle between lines drawn from any point on a hyperbola to the foci is a tangent.

Cor. 2.—The tangent at the vertex is perpendicular to the axis.

Cor. 3.—The normal to a hyperbola at any point bisects the external angle between the focal radii vectores drawn to the point.

PROPOSITION IV.

The locus of the foot of the perpendicular let fall from either focus on any tangent to a hyperbola is a circle described on the transverse axis as diameter.

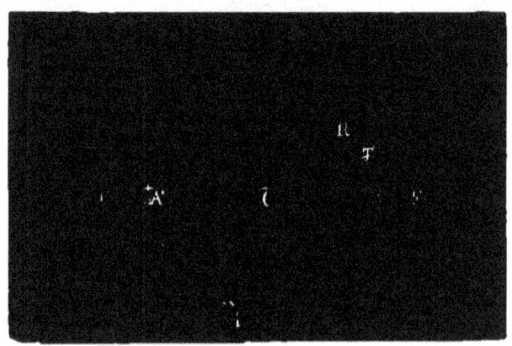

Fig. 6.

Let PT be any tangent, and FT a perpendicular let fall from the focus.

Produce FT to meet $F'P$ in R; join TC.

Then the $\triangle^s FTP$ and RTP are equal;   (26, I. Euclid.)

$$\therefore FT = TR.$$

Now, in $\triangle F'FR$, $F'F$ is bisected in C, and FR in T;

The Hyperbola.

$$\therefore CT = \tfrac{1}{2} FR = \tfrac{1}{2}(F'P - PR)$$
$$= \tfrac{1}{2}(F'P - PF)$$
$$= \tfrac{1}{2} AA' \qquad \text{(Prop. ii.)}$$
$$= CA.$$

Hence the locus of T is a circle described with the centre C and radius CA.

Cor. 1.—If the vertex of a right angle FTL move along a fixed circle, while one leg passes through a fixed point F without that circle, the other leg will always touch a hyperbola.

Draw the diameter of the circle passing through F; cut off $CF' = CF$; produce FT till $TR = TF$; join $F'R$, and produce it to meet TL in P; then P is the point at which TL will touch the hyperbola.

Join FP, CT.

Then the $\triangle^s FTP$ and RTP are equal; (4, I. Euclid.)

$$\therefore FP = RP.$$

Also, since $FT = TR$ and $FC = CF'$;

$$\therefore CT \parallel F'R \text{ and } = \tfrac{1}{2} F'R;$$
$$\therefore PF' - PF = PF' - PR$$
$$= F'R$$
$$= 2CT$$
$$= 2CA;$$

hence P is a point on the hyperbola of which F and F' are the foci, and AA' the transverse axis.

Cor. 2.—If FT become a tangent to the circle, then CT will be $\perp FT$, and $\therefore$ in directum with TL; and $\therefore F'R$, which is $\parallel CT$, will meet TL at infinity; hence CL is an asymptote to the curve.

Fig. 7.

If FT' be the other tangent to the circle drawn from F, then CT' will also be an asymptote.

Cor. 3.—If TO be drawn $\perp$ the axis, then TO is the directrix.

For $$CF \times CO = CA^2;\qquad (8,\ \text{VI. Euclid.})$$

$\therefore O$ is a point on the directrix. (*Cor.* 4, Prop. I.)

Cor. 4.—If AH be drawn $\perp CF$, meeting CL in H, and HB be drawn $\parallel CF$, meeting CB drawn $\perp CF$ in B, then CB is the *semi-conjugate axis*.

For the $\triangle^s CFT$ and CHA are equal; (26, I. Euclid.)

$$\therefore TF = AH = CB;$$

$$\therefore CB^2 = FT^2 = FA \times FA'.$$

(36, III. Euclid.)

Hence the point B, together with the corresponding point B', obviously correspond to the extremities of the axis minor in the Ellipse. (See *Cor.* 3, Prop. II., page 51.)

Cor. 5. $CF^2 = CA^2 + CB^2.$

For $CF^2 = CH^2 = CA^2 + AH^2 = CA^2 + CB^2.$

Cor. 6.—If we cut off Cf and Cf' each $= CF$ or CH, and consider f, f' the foci, and BB' the transverse axis of a hyperbola, then AA' will be its conjugate axis.

For the square of its semi-conjugate axis

$$= Cf^2 - CB^2 \qquad (Cor.\ 5.)$$

$$= CH^2 - AH^2$$

$$= CA^2.$$

The hyperbola described with BB' as the transverse axis, and f, f' as foci, is called the hyperbola *conjugate* to the original hyperbola.

Cor. 7.—The lines joining the extremities of the axes are bisected by one asymptote and parallel to the other.

For $CAHB$ is a rectangle; $\therefore AB$ is bisected by CH.

Also, $CB = AH = AH'$ and $CB \parallel AH'$;

$$\therefore AB \parallel CH'.$$

(33, I. Euclid.)

Cor. 8.—The perpendicular from the focus on either asymptote is equal to the semi-conjugate axis.

For $\triangle CAH = \triangle CTF;$ (26, I. Euclid.)

$$\therefore FT = AH = CB.$$

NOTE.—When *the axis* only is mentioned it is always understood to be the transverse axis.

Proposition V.

The distance of any point P on the curve from the focus F is equal to the length of the line Pq drawn from the point parallel to an asymptote to meet the directrix.

Fig. 8.

On the transverse axis as diameter describe a circle. Draw the tangent FT; join CT, and draw $TO \perp CF$. Then OT is the directrix, and CT an asymptote.

(Cors. 2 & 3, Prop. IV.)

Also $\qquad Pq : Pp = CT : CO \qquad$ (Similar $\triangle^s$.)
$$= CA : CO$$
$$= PF : Pp. \quad (Cor. 3, Prop. I.)$$

Hence, $\qquad PF = Pq.$

Cor.—Hence, being given the eccentricity of a hyperbola, we can find the angle between the asymptotes.

For $\qquad CT : CO = Pq : Pp \qquad$ (Similar $\triangle^s$.)
$$= PF : Pp$$
$$= \epsilon : 1.$$

Now in the $\triangle\ CTO$ we know $\angle$ at $O = 90°$; also the ratio of the sides $(CT : CO)$; hence the $\angle\ TCO$, or half the $\angle$ between the asymptotes, is known.

Proposition VI.

The rectangle under the focal perpendiculars on any tangent to a hyperbola is equal to the square of the semi-conjugate axis.

Fig. 9.

Let FT, $F'T'$, be perpendiculars from the foci on the tangent at any point P.

Join TC, and produce it to meet $F'T'$ in S.

Then the $\triangle^s FCT$ and $F'CS$ are equal;
(26, I. Euclid.)

$\therefore F'S = FT$, and $CS = CT = CA$; (Prop. iv.)

$\therefore S$ is a point on the auxiliary circle;

$\therefore A'F' \times F'A = F'T' \times F'S$ (35, III. Euclid.)

$= F'T' \times FT$;

$\therefore F'C^2 - CA'^2 = F'T' \times FT$, (5, II. Euclid.)

or $\qquad BC^2 = F'T' \times FT.$ (Cor. 5, Prop. iv.)

Proposition VII.

The locus of the intersection of tangents to a hyperbola which cut at right angles is a circle.

Fig. 10.

Describe the auxiliary circle cutting the tangent QP in T and T', and the tangent QP' in H and H'. Join FH, FT, $F'H'$, $F'T'$, QC; produce QC to meet the auxiliary circle in K and L.

Then $FT, F'T'$, are $\perp QP$, and $\therefore \parallel QP'$;
(Prop. iv.)

and $FH, F'H' \perp QP'$; (Prop. iv.)

hence $F'T' = QH'$, and $FT = QH$. (34, I. Euclid.)

Now $CQ^2 = CL^2 - KQ \times QL$ (5, II. Euclid.)

$= CL^2 - QH' \times QH$ (35, III. Euclid.)

$= CA^2 - FT \times F'T'$

$= CA^2 - BC^2.$ (Prop. v.)

Hence the locus of Q is a circle described with the centre C, and is called *the director circle* of the hyperbola.

Proposition VIII.

If PN be a normal at any point P on a hyperbola, and PM an ordinate to the transverse axis; then

$$CA^2 : CF^2 :: CM : CN.$$

Fig. 11.

Join PF, PF'; draw Ppp' ∥ the axis, meeting the directrices in p, p'.

Since PN bisects the external ∠ between PF and PF';
(*Cor.* 3, Prop. III.)

$$\therefore F'N : NF = PF' : PF$$
(Prop. B., VI. Euclid.)
$$= Pp' : Pp.$$

Comp. $F'N + NF : F'N - NF = Pp' + Pp : Pp' - Pp$;

$$\therefore 2CN : 2CF = 2CM : 2CO.$$

Alt. $\quad CN : CM = CF : CO$
$$= CF^2 : CF \times CO$$
$$= CF^2 : CA^2.$$
(*Cor.* 4, Prop. I.)

Cor. $\quad CM : MN = CA^2 : CB^2.$

For $\quad CM : CN - CM = CA^2 : CF^2 - CA^2$;

or $\quad CM : MN = CA^2 : CB^2.$
(*Cor.* 4, Prop. IV.)

Proposition IX.

If a tangent at any point P of a hyperbola be produced to meet the axis in T, and the ordinate PM be drawn; then CA will be a mean proportional between CM and CT.

Fig. 12.

Through P draw Ppp' ∥ to the axis; join PF, PF'.

Then, since PT is the bisector of the $\angle F'PF$;
(Cor. 1, Prop. III.)
$$\therefore F'T : TF = F'P : PF \quad \text{(3, VI. Euclid.)}$$
$$= Pp' : Pp.$$

Comp. and div.
$$F'T + TF : F'T - TF = Pp' + Pp : Pp' - Pp;$$
or
$$2CF : 2CT = 2CM : 2CO;$$
$$\therefore CT \times CM = CF \times CO$$
$$= CA^2. \quad \text{(Cor. 4, Prop. I.)}$$

Cor. 1.—If the auxiliary circle be described, and TQ drawn $\perp$ the axis; then QM is a tangent to the circle. Join CQ.

Since
$$CT : CA = CA : CM;$$
$$\therefore CT : CQ = CQ : CM;$$
$\therefore$ the $\triangle^s$ CQT and CMQ are similar;
(6, VI. Euclid.)
$\therefore \angle CQM = \angle CTQ$; $\therefore QM$ is a tangent.

CHAP. III.]   *The Hyperbola.*   127

Cor. 2.—If from the foot of any ordinate of a hyperbola a tangent be drawn to the auxiliary circle, this tangent is to the ordinate in a constant ratio.

Draw the normal PN.

Then, since the $\triangle TPN$ is right-angled, and $PM \perp TN$;

$$\therefore TM \times MN = PM^2. \qquad (8, \text{VI. Euclid.})$$

Also, since $\triangle CQM$ is right-angled, and $QT \perp CM$;

$$\therefore CM \times TM = QM^2.$$

Hence $\qquad QM^2 : PM^2 = CM : MN$

$$= CA^2 : CB^2;$$

(*Cor.*, Prop. VIII.)

$$\therefore QM : PM = CA : CB.$$

Cor. 3.—If QA be drawn and produced to meet PM produced in K, then

$$MK = MQ, \text{ and } PM : MK = CB : CA.$$

For $\angle MQA = \angle QA'A$ (32, III. Euclid.)

$\qquad = \angle AQT$ (13, VI. Euclid.)

$\qquad = \angle AKM;$

$\therefore MK = MQ;$

Also $PM : QM = CB : CA;$

$\therefore PM : MK = CB : CA.$

Cor. 4.—The curve is concave towards the transverse axis.

(See Fig. 13.)

Draw any line perpendicular to the axis between the vertex A and the point P; let it meet the curve in p, AP in S, the axis in m, and QA produced in r. From m draw the tangent mq to the auxiliary circle; draw $qn \perp$ to the axis; join np; produce qA to meet pr produced in k.

Then $\qquad PM : MK = CB : CA = pm : mk;$ (*Cor.* 2.)

but $\qquad PM : MK = sm : rm;$ (Similar $\triangle^s$.)

$\therefore pm : mk = sm : rm.$

Alt. $\qquad pm : sm = mk : rm$

$\qquad\qquad = qn : vn;$

but qn is $> vn$ by property of the circle;

$\therefore pm$ is $> sm.$

That is, if a perpendicular be drawn to the transverse axis produced, between the vertex A and any point P on the branch of the curve of which A is the vertex, the segment intercepted on it between the curve and the transverse axis is greater than the segment intercepted on it between the line AP and the transverse axis; and, therefore, the curve is concave towards the transverse axis.

[Chap. III.] *The Hyperbola.* 129

Proposition X.

The rectangle under the segments of the axis made by any ordinate is to the square of the ordinate in a constant ratio.

See Fig. 12.

For $PM^2 : QM^2 = CB^2 : CA^2$; (*Cor.* 2, Prop. IX.)
but $QM^2 = MA \times MA'$; (36, III. Euclid.)
$\therefore PM^2 : MA \times MA' = CB^2 : CA^2$.

Cor. 1.—The Latus rectum is a third proportional to the transverse and conjugate axes.

$$\left(\frac{L}{2}\right)^2 : AF \times FA' = CB^2 : CA^2;$$

$$\therefore \left(\frac{L}{2}\right)^2 : CB^2 = CB^2 : CA^2.$$

Hence $CA : CB = CB : \dfrac{L}{2}$. (*Cor.* 4, Prop. IV.)

Cor. 2. $CB^2 = CM' \times CT'$.

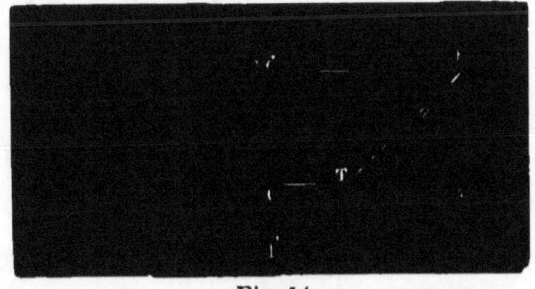

Fig. 14.

For $CM^2 : CA^2 = CM : CT$; (Prop. IX.)
$\therefore CM^2 - CA^2 : CA^2 = CM - CT : CT$.
$PM^2 : CB^2 = TM : CT$ (Prop. X.)
$= PM : CT$ (Similar $\triangle$'s.)
$= PM^2 : PM \times CT'$;
$\therefore CB^2 = CM' \times CT'$.

K

Proposition XI.

The locus of the intersection of any tangent to a hyperbola, with the line drawn through the focus perpendicular to the radius vector drawn from the focus to the point of contact of the tangent, is the directrix.

Fig. 15.

Let PT be any tangent, and $FT \perp FP$.

Draw $TH \perp PF'$, and $TO \perp$ the axis.

The $\triangle^s PFT$ and PHT are equal.   (26, I. Euclid.)

$\therefore FT = TH$, and $PH = PF$;

$\therefore F'H = F'P - PH = F'P - PF = AA'$.
(Prop. 11.)

Now $\quad F'H^2 = F'T^2 - TH^2 \quad$ (47, I. Euclid.)

$\qquad = F'T^2 - TF^2$

$\qquad = F'O^2 - OF^2$;

$\therefore AA'^2 = 4CF \times CO$;   (Cor. 5, II. Euclid.)

$\therefore CA^2 = CF \times CO$.

Hence OT is the directrix.   (Cor. 4, Prop. I.)

Cor.—Conversely, if from any point on the directrix a tangent be drawn to a hyperbola, the line joining that point to the focus is perpendicular to the radius vector drawn from the focus to the point of contact of the tangent.

Proposition XII.

If any chord PP' of a hyperbola cut the directrix in D, and if F be the focus corresponding to the directrix, on which D is situated; then FD is the external bisector of the $\angle PFP'$.

Fig. 16.

Draw Pp, $P'p' \perp$ the directrix; produce $P'F$.

Then $\qquad PF : Pp = P'F : P'p'$.

Alt. $\qquad PF : P'F = Pp : P'p'$

$\qquad\qquad\qquad = PD : P'D;$ (Similar $\triangle^s$.)

$\therefore FD$ bisects the $\angle PFQ$. (Prop. B., VI. Euclid.)

Cor. 1.—Hence, being given one focus F, and three points P, P', P'', on a hyperbola, we can find the directrix and axes.

For, draw FD bisecting the angle between PF and $P'F$ produced; the point where this line meets $P'P$ produced will be one point on the directrix. Similarly, by bisecting the angle between PF and $P''F$ produced, another point on the directrix may be found; and hence the directrix and axes.

If a tangent be defined as the line joining two indefinitely near points on the curve, it will follow immediately from the proposition that the right line drawn from the focus to the point of intersection of any tangent with the directrix is perpendicular to the radius vector drawn from the focus to the point of contact of the tangent; also, that the tangent bisects the angle between the radii vectores drawn from the foci to the point of contact.

For when P is indefinitely near to P', PD becomes a tangent at the point P, and the $\angle PFP'$ is indefinitely small (see Fig. 16);

$$\therefore \angle PFQ = 180°,$$

but $\angle PFD = \tfrac{1}{2} \angle PFQ$;

$$\therefore \angle PFD = 90°.$$

Also (see Fig. 17),

Fig. 17.

$$PF : PF' = Pp : Pp'$$
$$= PD : PD', \quad \text{(Similar } \triangle^s\text{.)}$$

and $\angle PFD = PF'D'$, each being $90°$;

$$\therefore \triangle PFD \text{ is similar to } \triangle PF'D';$$

hence $\angle FPD = \angle F'PD'$.

Proposition XIII.

If two fixed points, P, P', on a hyperbola be joined with a third variable point O, the segment pp' intercepted on either directrix by the produced chords subtend a constant angle at the focus corresponding to that directrix.

Fig. 18.

Since Fp is the external bisector of the $\angle PFO$;

(Prop. xii.)

$$\therefore \angle OFp + \tfrac{1}{2} \angle OFP = 90°.$$

Also Fp' is the external bisector of the $\angle OFP'$;

(Prop. xii.)

$$\therefore \angle OFp' + \tfrac{1}{2} \angle OFP' = 90°;$$

$$\therefore \angle OFp + \tfrac{1}{2} \angle OFP = \angle OFp' + \tfrac{1}{2} \angle OFP';$$

$$\therefore \angle OFp - \angle OFp' = \tfrac{1}{2}(\angle OFP' - \angle OFP)$$

or $\angle pFp' = \tfrac{1}{2} \angle PFP'$, and $\therefore$ constant.

Cor.—The anharmonic ratio of the pencil formed by joining four fixed points on a hyperbola to any fifth variable point is constant.

For $\quad O \cdot PP'P''P''' = O \cdot pp'p''p''' = F \cdot pp'p''p'''.$

134 The Hyperbola. [CHAP. III.

PROPOSITION XIV.

The line joining the focus to the intersection of two tangents to a hyperbola bisects the angle which the points of contact subtend at the focus.

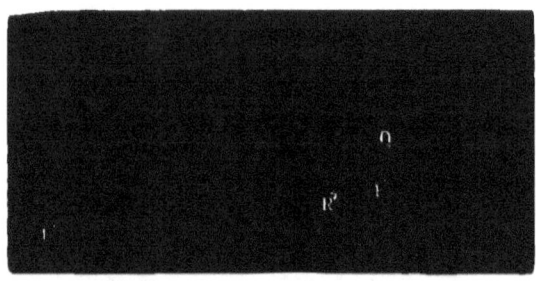

Fig. 19.

Let QP, QP' be tangents; join PF, PF', $P'F$, $P'F'$. Cut off $PR = PF$, and $P'R' = P'F$; join Q with R, R', F, F'.

Then the $\triangle^s QFP$ and QRP are equal;
(4, I. Euclid.)
$$\therefore QR = QF.$$

Again, the $\triangle^s QFP'$ and $QR'P'$ are equal;
(4, I. Euclid.)
$$\therefore QR' = QF;$$
hence $\quad QR = QR';$
but $\quad F'R = F'P - PR$
$\quad\quad = F'P - PF$
$\quad\quad = P'F' - P'F$ $\quad$ (Prop. XI.)
$\quad\quad = P'F' - P'R'$
$\quad\quad = F'R'.$

Hence the $\triangle^s RF'Q$ and $R'F'Q$ are equal;
(8, I. Euclid.)
$$\therefore \angle RF'Q = \angle R'F'Q.$$

[Chap. III.] *The Hyperbola.* 135

Cor.—It is obvious from the above demonstration, also, that
$$\angle PFQ = \angle PRQ = \angle QR'P'$$
$$= \angle QFP'.$$

Proposition XV

The angle subtended at either focus by the segment intercepted on a variable tangent to a hyperbola by two fixed tangents is constant.

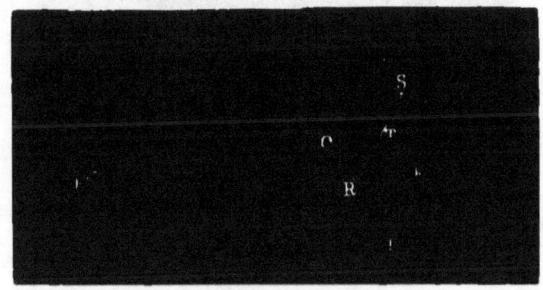

Fig. 20.

Let QP, QP' be the two fixed tangents, and RS the variable tangent.

Join FP, FS, FT, FR, FP'.

Then $\angle TFS = \tfrac{1}{2} \angle TFP$; also $\angle TFR = \tfrac{1}{2} \angle TFP'$;
(Prop. xiii.)

$\therefore \angle TFS + \angle TFR = \tfrac{1}{2} \angle TFP + \tfrac{1}{2} \angle TFP'$,

or $\angle RFS =$ a constant.

Cor.—The anharmonic ratio of the four points in which a variable tangent to a hyperbola is cut by four fixed tangents is constant.

For the segments intercepted on the variable tangent subtend constant angles at the focus.

Proposition XVI.

If a tangent at the extremity of any diameter meet the axis in *T*, the area of the triangle *CPT* thus formed will be equal to the area of the triangle *CAH*, formed by drawing a tangent at the vertex.

Fig. 21.

Draw the ordinate PM; join PA, TH.

Then $\quad\quad TC : CA = CA : CM \quad\quad$ (Prop. IX.)

$\quad\quad\quad\quad\quad\quad\quad = CH : CP ; \quad$ (Parallel lines.)

$\quad\quad \therefore PA$ is $\parallel HT;\quad\quad$ (2, VI. Euclid.)

$\quad\quad \therefore \triangle PHT = \triangle HTA ;\quad$ (37, I. Euclid.)

hence $\quad\quad \triangle CPT = \triangle CAH.$

Cor. $\quad\quad \triangle PMT =$ trap. $PMAH.$

If we take the equal $\triangle^s\ CPT$ and CAH, in succession from the $\triangle CPM$;

then $\quad\quad \triangle PMT =$ trap. $PMAH.$

Proposition XVII.

If an ordinate to the axis drawn from any point P on a hyperbola be produced to meet the asymptotes in D and D', then $PD \times PD'$ = square of semi-conjugate axis.

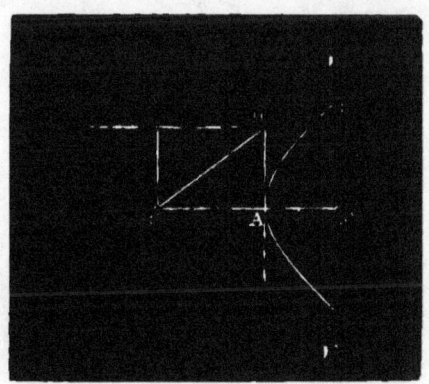

Fig. 22.

For $\quad MD^2 : MC^2 = AH^2 \text{ or } BC^2 : AC^2;$
$\quad\quad\quad\quad\quad\quad\quad\quad\quad\quad\quad\quad$ (Similar $\triangle^s$.)

but $\quad PM^2 : MC^2 - CA^2 = BC^2 : CA^2;\quad$ (Prop. x.)

$\therefore\ MD^2 - PM^2 : CA^2 = BC^2 : CA^2;$

hence $\quad MD^2 - PM^2 = BC^2,$

or $\quad\quad PD \times PD' = BC^2.\quad$ (5, II. Euclid.)

Note.—In the figures connected with some of the following Propositions, in order to avoid confusion, occasionally lines are not actually drawn: when a letter is omitted, reference to any preceding figure, where the same letter has been used, will at once indicate the point.

138　　　　　*The Hyperbola.*　　　[CHAP. III.

Proposition XVIII.

The segments PQ, $P'Q'$ intercepted on *any* line between the curve and its asymptotes are equal.

Fig. 23.

Through P and P' draw $\perp^s$ to the axis.

Then $\quad QP : PR = QP' : P'S,\quad$ (Similar $\triangle^s$.)

and $\quad PQ' : Pr = Q'P' : P'S'$;

$\therefore\ QP \times PQ' : PR \times Pr = QP' \times Q'P' : P'S \times P'S'$;

but $\quad PR \times Pr = BC^2 = P'S \times P'S'\quad$ (Prop. XVII.)

$\therefore\ QP \times PQ' = QP' \times Q'P'$,

or $QP \times PP' + QP \times Q'P' = Q'P' \times PP' + Q'P' \times QP$;
$\qquad\qquad\qquad\qquad\qquad$ (1, II. Euclid.)

$\therefore\ QP \times PP' = Q'P' \times PP'$,

or $\qquad\qquad QP = Q'P'$.

Cor. 1.—If a line drawn from C through the middle point of QQ' meet the curve in D, then the line drawn through D parallel to QQ' will be a tangent.

For, suppose it meet the curve again in D', and let it meet the asymptotes in E and E'.

Then, by this Proposition, $DE = D'E'$.

But in the $\triangle QCQ'$, since QQ' is bisected in G, and $EE' \parallel QQ$;

$$\therefore DE = DE'.$$

Hence $DE' = D'E'$, which is impossible; $\therefore$ &c.

Cor. 2.—Any line PP' is an ordinate to the diameter passing through its middle point, and the portion of any tangent intercepted between the asymptotes is bisected at its point of contact.

Cor. 3.—If any line be drawn parallel to PP' or EE', it is obvious that the portion intercepted between the asymptotes will be bisected by the diameter CD; but the segments intercepted between the curve and the asymptotes are also equal, and hence *any diameter bisects all its ordinates.*

Cor. 4. $PQ \times PQ' = DE^2.$

For, if through D a perpendicular be drawn to the axis, meeting the asymptotes in K, K'.

Then $DK : DE = PR : PQ$; (Similar $\triangle^s$.)

also $DK' : DE' = Pr : PQ'$; (Do.)

$\therefore DK \times DK' : DE \times DE' = PR \times Pr : PQ \times PQ'$;

but $DK \times DK' = BC^2 = PR \times Pr$; (Prop. xvii.)

$\therefore PQ \times PQ' = DE \times DE'$

$\qquad\qquad\quad = DE^2.$ (See *Cor.* 1.)

Cor. 5.—If through any point on a hyperbola a line be drawn in a constant direction, the rectangle under the segments intercepted between the point and the asymptotes is constant.

140   *The Hyperbola.*   [CHAP. III.

PROPOSITION XIX.

If CP' be the semi-diameter conjugate to CP, and tangents be drawn at P, P', meeting the transverse and conjugate axes in T, T' respectively; TQ, $T'Q'$ perpendiculars to the axes meeting the circles described on the axes as diameters in Q, Q'; then $\angle QCA = \angle Q'CB$.

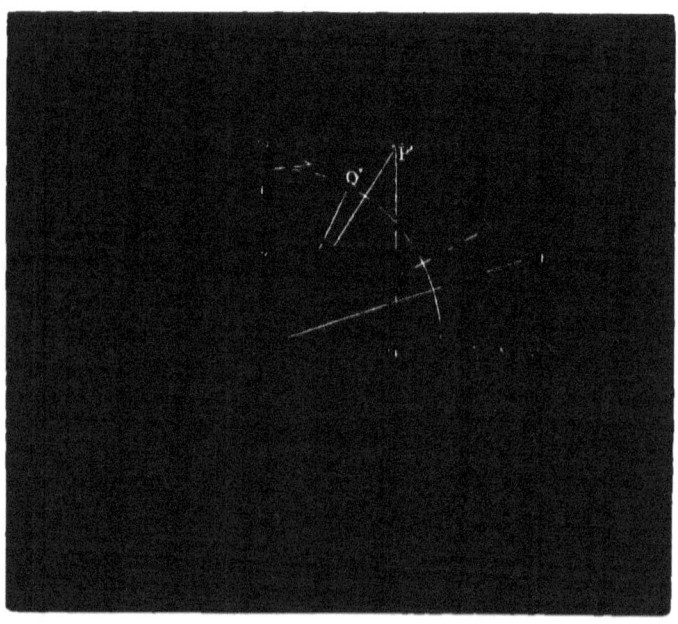

Fig. 24.

Join QM, $Q'm'$.

Then QM, $Q'm'$ are tangents to the circles at Q, Q', respectively.   (*Cor.* 1, Prop. IX.)

$$PM : MQ = CB : CA, \quad (Cor.\ 2,\ \text{Prop. IX.})$$
and
$$P'm' : m'Q' = CA : CB;$$
$$\therefore\ P'm' \times PM = Q'm' \times QM.$$

CHAP. III.]   *The Hyperbola.*   141

But since CP' is $\parallel PT$;

$$\therefore PM : MT = Cm' : m'P';$$
$$\therefore PM \times m'P' = Cm' \times MT;$$
$$\therefore Q'm' \times QM = Cm' \times MT;$$
$$\therefore Cm' : m'Q' = QM : MT$$
$$= CM : MQ;$$
(8, VI. Euclid.)

hence the $\triangle^s\ m'CQ'$ and MCQ are similar;
(7, VI. Euclid.)

and $\therefore \angle QCA = \angle Q'CB$.

Cor. 1.—CP is parallel to the tangent at P'.

For $P'm' : m'T' = \begin{cases} P'm' : m'Q' \\ m'Q' : m'T' \end{cases}$

$= \begin{cases} CA : CB \\ Cm' : m'Q' \end{cases}$ (*Cor.* 2, Prop. IX.)
(8, VI. Euclid.)

$= \begin{cases} QM : MP \\ CM : QM \end{cases}$ (*Cor.* 2, Prop. IX.)
(Prop. XIX.)

$= CM : MP = Pm : Cm;$

$\therefore \triangle^s\ P'm'T'$ and CPm are similar, and $\therefore CP \parallel P'T''$.

Hence, if one diameter CP' of an hyperbola be conjugate to another CP, then, conversely, CP will be conjugate to CP'.

Cor. 2. $PM = Q'm' = Cm$; and $P'm' = QM = CM'$.

Since the $\triangle^s\ Cm'Q'$ and CMQ are similar,

$$m'Q' : MQ = CQ' : CQ$$
$$= CB : CA.$$

But $PM : MQ = CB : CA;$ (*Cor.* 2, Prop. IX.)
$\therefore m'Q' : MQ = PM : MQ;$
$\therefore m'Q' = PM.$

Also $P'm' : m'Q' = CA : CB$
$= QM : Q'm';$ (Similar $\triangle^s$.)
$\therefore P'm' = QM.$

Cor. 3. $\qquad CM^2 - CM'^2 = CA^2.$

For $CA^2 = CQ^2 = CM^2 - QM^2 = CM^2 - CM'^2.$ (Cor. 2.)

Cor. 4. $\qquad P'M'^2 - PM^2 = CB^2.$

For $CB^2 = CQ'^2 = Cm'^2 - m'Q'^2 = PM'^2 - PM^2.$ (Cor. 2.)

Cor. 5.—The difference of the squares of any pair of conjugate semi-diameters is equal to the difference of the squares of the semiaxes.

For
$$CA^2 - CB^2 = CM^2 - CM'^2 - (P'M'^2 - PM^2)$$
$$(Cors.\ 3\ \text{and}\ 4.)$$
$$= CM^2 + PM^2 - (CM'^2 - P'M'^2)$$
$$= CP^2 - CP'^2.$$

Cor. 6. $\quad PM : CM' = CB : CA = P'M' : CM.$

For $\qquad PM : QM = CB : CA;\quad$ (Cor. 2, Prop. ix.)

but $\qquad QM = m'P' = CM';$

$\therefore PM : CM' = CB : CA.$

Again, $\quad Cm' : CM = CQ' : CQ;\qquad$ (Similar $\triangle^s$.)

$\therefore P'M' : CM = CB : CA.$

Cor. 7.—The $\triangle^s$ CPM and $CP'M'$ are equal in area.

For $\qquad PM \times CM = P'M' \times CM'.$ (Cor. 6.)

Cor. 8. Mm' is parallel to an asymptote.

For $\qquad Cm' : Cm = CQ' : CQ\qquad$ (Similar $\triangle^s$.)
$$= CB : CA;$$

$\therefore Mm'$ is $\parallel AB$, and $\therefore \parallel$ an asymptote.
(Cor. 7, Prop. iv.)

Cor. 9.—If MP, $m'P'$ be produced to meet in O, then CO will be an asymptote.

Since $CMOm'$ is a parallelogram, CO will bisect Mm', and $\therefore$ it will bisect AB, which is $\parallel Mm'$.

Hence CO is an asymptote. $\qquad$ (Cor. 7, Prop. iv.)

Proposition XX.

If CP, CP' be conjugate semi-diameters, PM, PM' ordinates to the axis, PT a tangent at P; then

$$\triangle CVM' = \triangle PTM, \text{ and } \triangle CP'V = \triangle CPT.$$

Fig. 25.

For $\triangle CPM : \triangle CHA : \triangle CVM' = CM^2 : CA^2 : CM'^2$;
(19, VI. Euclid.)

$\therefore \triangle CPM - \triangle CHA : \triangle CVM' = CM^2 - CA^2 : CM'^2$;

but $\qquad CM^2 - CA^2 = CM'^2$; (Cor. 3, Prop. XIX.)

$\therefore \triangle CPM - \triangle CHA = \triangle CVM'$,

or $\qquad$ trap. $PMAH = \triangle CVM'$;

$\therefore \triangle PTM = \triangle CVM'$.
(Cor., Prop. XVI.)

Again, $\qquad \triangle CPM = \triangle CP'M'$;
(Cor. 7, Prop. XIX.)

$\therefore \triangle CPM - \triangle PTM = \triangle CP'M' - \triangle CVM'$;

$\therefore \triangle CP'V = \triangle CPT$.

144 · The Hyperbola. [CHAP. III.

Proposition XXI.

If QG be an ordinate drawn to any diameter Pp, from any point Q on a hyperbola, cutting the transverse axis in L; QS an ordinate drawn to the axis from the same point and produced, if necessary, to meet the diameter Pp in K; AH a tangent at the vertex; then $\triangle\, QLS$ = trap. $KSAH$.

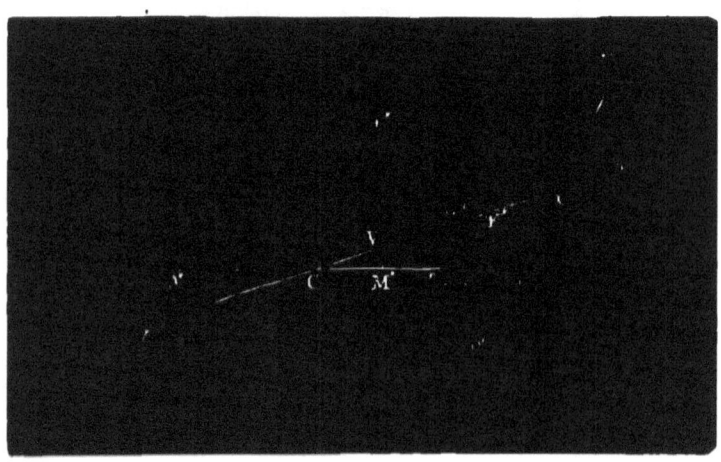

Fig. 26.

Let CP' be the diameter conjugate to CP; draw the ordinates PM, $P'M'$ to the axis.

Then $\quad CA : CS : CM = AH : SK : MP$;

(Similar $\triangle^s$.)

$\therefore\; CA + CS : CA + CM = AH + SK : AH + MP$,

or $\quad A'S : A'M = AH + SK : AH + MP$;

$\therefore A'S \times SA : A'M \times MA = (AH + SK)\, SA : (AH + MP)\, MA$;

$\therefore\; QS^2 : PM^2$ = trap. $KSAH$: trap. $PMAH$;

(Prop. x.)

$\therefore \triangle QSL : \triangle PMT$ = trap. $KSAH$: trap. $PMAH$,
(19, VI. Euclid.)
but $\triangle PMT$ = trap. $PMAH$; (Cor. Prop. xvi.)

$\therefore \triangle QSL$ = trap. $KSAH$.

Cor. 1. $\triangle QGK$ = trap. $TPGL$.

For $\triangle QSL$ = trap. $KSAH$.

$\therefore \triangle QGK$ = trap. $GLAH$ = trap. $TPGL$.
(Prop. xvi.)

Cor. 2.—In like manner it can be shown that

$$\triangle GQ'K' = \text{trap. } TPGL;$$

$$\therefore \triangle GQK = \triangle GQ'K';$$

but they are also similar; $\therefore GQ = GQ'$.

Hence, any diameter bisects all chords parallel to the tangents at its extremities. (See also *Cor.* 3, Prop. xviii.)

Proposition XXII.

If CP' be the semi-diameter conjugate to CP, QG any ordinate to CP; then $QG^2 : PG \times Gp = CP'^2 : CP^2$. (See Fig. 26.)

For $\triangle CPT : \triangle CGL = CP^2 : CG^2$.
(19, VI. Euclid.)

Division, $\triangle CPT$: trap. $TPGL = CP^2 : CG^2 - CP^2$,

or $\triangle CP'V : \triangle QGK = CP^2 : PG \times Gp$;
(Prop. xx.)

$\therefore CP'^2 : QG^2 = CP^2 : PG \times Gp$;
(19, VI. Euclid.)

hence $QG^2 : PG \times Gp = CP'^2 : CP^2$.

L.

146   The Hyperbola.   [CHAP. III.

Cor. 1.—Ordinates to any diameter at equal distances from the centre are equal.

For the rectangles are the same for equal distances from the centre.

Cor. 2.— $QG^2 : CG^2 - CP^2 = CP'^2 : CP^2$.

For $PG \times Gp = CG^2 - CP^2$.

Cor. 3.—If the ordinate GQ be produced to meet the conjugate hyperbola in Q'; then

$$Q'G^2 : CG^2 + CP^2 = CP'^2 : CP^2.$$

Fig. 27.

Draw $Q'G'$ an ordinate to CP'.

Then $Q'G'^2 : CG'^2 - CP'^2 = CP^2 : CP'^2$.   (*Cor.* 2.)

Alt. $CG^2 : CP^2 = CG'^2 - CP'^2 : CP'^2$;

∴ $CG^2 + CP^2 : CP^2 = CG'^2 : CP'^2$;

∴ $Q'G^2 : CG^2 + CP^2 = CP'^2 : CP^2$.

Cor. 4.—The squares of the ordinates to any diameter are proportional to the rectangles under the segments of that diameter.

CHAP. III.]   *The Hyperbola.*   147

PROPOSITION XXIII.

If a normal at any point P of a hyperbola meet the transverse axis in N, and the conjugate axis in N'; then
$$PN : CP' = CB : CA = CP' : PN',$$
CP' being the semi-diameter conjugate to CP.

Fig. 28.

Since CP' is $\parallel PT$, the $\triangle P'Cm'$ is similar to $\triangle PMT$, and $\therefore$ similar to $\triangle NPM$;

$$\therefore P'C : NP = P'm' : PM$$
$$= MQ : PM \quad (Cor.\ 2,\ \text{Prop. XIX.})$$
$$= CA : CB. \quad (Cor.\ 2,\ \text{Prop. IX.})$$

Also $\quad PN' : PN = CM : MN \quad$ (Parallel lines.)
$$= CA^2 : CB^2; \quad (Cor.,\ \text{Prop. VIII.})$$

$\therefore PN' \times PN : PN^2 = CA^2 : CB^2 = P'C^2 : PN^2;$

hence $\quad PN \times PN' = P'C^2,$

or $\quad PN' : P'C = P'C : PN = CA : CB.$

Cor.— $\quad PN \times PN' = P'C^2.$

148   *The Hyperbola.*   [CHAP. III.

PROPOSITION XXIV.

The rectangle under the distances of the foci, from any point P on a hyperbola, is equal to the square of the semi-diameter CP' conjugate to that passing through the point.

Fig. 29.

Circumscribe the $\triangle FPF'$ by a circle, cutting the conjugate axis in L and N'; join PN', FN'; produce PN' to meet the transverse axis in N.

Then, since LN' bisects FF', and is also perpendicular to it; ∴ LN' is the diameter of the circle, and arc FN' = arc $F'N'$;

hence $\angle F'PL = FPL$; ∴ PL is a tangent.   (Prop. III.)

Also $\angle LPN' = 90°$; ∴ PN is a normal.

Now $\angle FPN = \angle F'PN'$, and $\angle PFN = \angle PN'F$;
(26, III. Euclid.)

∴ the $\triangle^s FPN$ and $F'PN'$ are similar;

∴ $F'P : PN' = PN : PF$;

∴ $PF \times PF' = PN \times PN' = CP'^2$.
(Cor., Prop. XXIII.)

CHAP. III.]   *The Hyperbola.*   149

Otherwise thus—
$$2CA = PF' - PF;$$
$$\therefore 4CA^2 = PF'^2 + PF^2 - 2PF' \times PF$$
(4, II. Euclid.)
$$= 2CP^2 + 2CF^2 - 2PF' \times PF;$$
(*Cor.*, 13, II. Euclid.)
$$\therefore PF' \times PF = CF^2 + CP^2 - 2CA^2$$
$$= CF^2 - CA^2 + CP^2 - CA^2$$
$$= CB^2 + CP^2 - CA^2 \ (Cor.\ 5,\ \text{Prop. IV.})$$
$$= CP^2 - CP^2 + CP'^2 \ (Cor.\ 5,\ \text{Prop. XIX.})$$
$$= CP'^2.$$

Cor. 1.—Since PL bisects the $\angle FPF'$, it is a tangent, and $\therefore PN$, which is $\perp PL$, is a normal. Hence the circle which passes through the foci and any point P on a hyperbola, passes also through the points in which the tangent and normal at P intersect the conjugate axis.

Cor. 2.—If K be the foot of the perpendicular let fall from N' on either PF' or PF produced; then
$$PK = \tfrac{1}{2}(PF' - PF) = \text{semi-transverse axis.}$$

Cor. 3.—If K' be the foot of the perpendicular let fall from L on PF'; then
$$F'K' = \tfrac{1}{2}(PF' - PF) = \text{semi-transverse axis.}$$

Cor. 4.— $CA : CB = CP' : PN.$

(See Fig. 11.)

For $F'P : PF = F'N : NF.$ (Prop. B., VI. Euclid.)
Div. $F'P - PF : PF = F'N - NF : NF.$
Alt. $2CA : 2CF = PF : NF.$
Similarly, $2CA : 2CF = PF' : NF';$
$$\therefore CA^2 : CF^2 = PF \times PF' : NF \times NF'.$$
$$\therefore CA^2 : CF^2 - CA^2 = PF \times PF' : NF \times NF' - PF \times PF';$$
$$\therefore CA^2 : CB^2 = CP'^2 : PN^2.$$

150   *The Hyperbola.*   [CHAP. III.

Proposition XXV.

If CK be a perpendicular let fall from the centre on the tangent to a hyperbola at any point P, and CP' the semi-diameter conjugate to CP; then $CK : CA = CB : CP'$.

Fig. 30.

Draw FT, $F'T' \perp$ the tangent at P.

Since the $\triangle^s FPT$ and $F'PT'$ are similar;

$$\therefore FP : F'P = FT : F'T' = F'P - FP : F'T' - FT$$
$$= CA : CK;$$
$$\therefore FP \times F'P : FT \times F'T' = CA^2 : CK^2,$$
but $\qquad FP \times F'P = CP'^2, \qquad$ (Prop. XXIV.)
and $\qquad FT \times F'T' = CB^2; \qquad$ (Prop. VI.)
$$\therefore CP'^2 : CB^2 = CA^2 : CK^2;$$
$$\therefore CP' : BC = CA : CK.$$

Cor.—The area of the triangle formed by joining the extremities of any pair of conjugate semi-diameters is constant.

For $\qquad CP' \times CK = CB \times CA$;

$\therefore$ area of $\triangle P'CP$ = area of $\triangle BCA$, and therefore constant.

[CHAP. III.] *The Hyperbola.* 151

Proposition XXVI.

If a chord PP' of a hyperbola pass through a fixed point O, the rectangle under its segments is to the square of the parallel semi-diameter CR in a constant ratio.

Fig. 31.

Draw the diameter passing through O, meeting the curve in S; draw the ordinates SM, SM'.

Then $PN^2 : CR^2 = CN^2 - CR'^2 : CR'^2$. (Prop. xxii.)

Comp. $PN^2 + CR^2 : CR^2 = CN^2 : CR'^2$

Alt. $PN^2 + CR^2 : CN^2 = CR^2 : CR'^2$.

Similarly, $SM^2 + CR^2 : CM^2 = CR^2 : CR'^2$;

∴ $PN^2 + CR^2 : SM^2 + CR^2 = CN^2 : CM^2$

$= ON^2 : SM^2$;

(Similar △^{s}.)

∴ $PN^2 + CR^2 - ON^2 : ON^2 = SM^2 + CR^2 - SM^2 : SM^2$;

∴ $CR^2 + PO \times OP' : ON^2 = CR^2 : SM^2$.

Alt. $CR^2 + PO \times OP' : CR^2 = ON^2 : SM^2$

$= CO^2 : CS^2$.

Division, $PO \times OP' : CR^2 = CO^2 - CS^2 : CS^2$.

152   *The Hyperbola.*   [CHAP. III.

CASE 2.—(The importance of this proposition renders it advisable to give a separate demonstration of each case).

As before, draw the diameter passing through O; also PN' and SM ordinates to the diameters CR and CR' respectively.

Fig. 32.

$$SM^2 : CM^2 - CR'^2 = CR^2 : CR'^2; \quad \text{(Prop. XXII.)}$$

$$\therefore SM^2 + CR^2 : CM^2 = CR^2 : CR'^2.$$

But $CN'^2 - CR^2 : PN'^2 = CR^2 : CR'^2;$ (Prop. XXII.)

$$\therefore SM^2 + CR^2 : CN'^2 - CR^2 = CM^2 : PN'^2,$$

or $\quad SM^2 + CR^2 : PN^2 - CR^2 = CM^2 : CN^2$

$$= SM^2 : ON^2;$$
(Similar △^{s}.)

$\therefore SM^2 + CR^2 - SM^2 : SM^2 = PN^2 - ON^2 - CR^2 : ON^2,$

or $\quad CR^2 : PO \times OP' - CR^2 = SM^2 : ON^2$

$$= CS^2 : CO^2;$$

$$\therefore PO \times OP' : CR^2 = CS^2 + CO^2 : CS^2.$$

[CHAP. III.] *The Hyperbola.* 153

CASE 3.—Same construction.

Fig. 33.

$$PN'^2 : CN'^2 - CR^2 = CR'^2 : CR^2. \quad \text{(Prop. XXII.)}$$

Also $\quad SM'^2 : CM'^2 - CR^2 = CR'^2 : CR^2; \quad \text{(Prop. XXII.)}$

$$\therefore CM'^2 - CR^2 : CN'^2 - CR^2 = SM'^2 : PN'^2,$$

or $\quad SM^2 - CR^2 : PN^2 - CR^2 = CM^2 : CN^2$

$$= SM^2 : ON^2;$$

$$\therefore SM^2 + CR^2 - SM^2 : ON^2 - PN^2 + CR^2 = SM^2 : ON^2,$$

or $\quad CR^2 : OP \times OP' + CR^2 = CS^2 : CO^2;$

(Similar $\triangle^s$.)

$$\therefore OP \times OP' : CR^2 = CO^2 - CS^2 : CS^2.$$

Cor. 1.—If two chords PP', pp' of a hyperbola intersect, the rectangles under the segments are proportional to the squares of the parallel semi-diameters.

Let Cr be the semi-diameter $\parallel pp'$.

Then $OP \times OP' : CR^2 = CO^2 - CS^2 : CS^2 = Op \times Op' : Cr^2;$

$$\therefore OP \times OP' : Op \times Op' = CR^2 : Cr^2.$$

154   *The Hyperbola.*   [CHAP. III.

Cor. 2.—If from any point two tangents be drawn to a hyperbola, the tangents will be proportional to the parallel semi-diameters.

Proposition XXVII.

If from a point without a hyperbola two tangents OP, OR be drawn, any line AQ' drawn parallel to either will be cut by the curve and chord of contact of the tangents in geometric proportion.

Fig. 34.

Draw the semi-diameters CE and $CD \parallel OP$ and OR respectively.

Then $AQ \times AQ' : AP^2 = CD^2 : CE^2$
(*Cor.* 1, Prop. XXVI.)

$= OR^2 : OP^2$
(*Cor.* 2, Prop. XXVI.)

$= AB^2 : AP^2$;   (Similar $\triangle^s$.)

$\therefore AQ \times AQ' = AB^2$.

Proposition XXVIII.

If a circle intersect a hyperbola in four points, the common chords will be equally inclined to the axis.

Fig. 35.

Let PP', QQ' be the common chords intersecting in O.

Draw the semi-diameters CR, $CS \parallel PP'$, QQ' respectively.

Then $OP \times OP' : OQ \times OQ' = CR^2 : CS^2$;
(*Cor*. 1, Prop. XXVI.)

but $OP \times OP' = OQ \times OQ'$; (35, III. Euclid.)

∴ $CR = CS$.

Hence CR and CS, and ∴ PP' and QQ', are equally inclined to the axis. (*Cor*. 1, Prop. I.)

Cor.—In like manner it can be shown that the common chords PQ and $P'Q'$, also PQ' and $P'Q$, are equally inclined to the axis.

Proposition XXIX.

If a tangent to a hyperbola at any point P meet any diameter AA' in T, and the ordinate PM be drawn; then CA is a mean proportional between CM and CT.

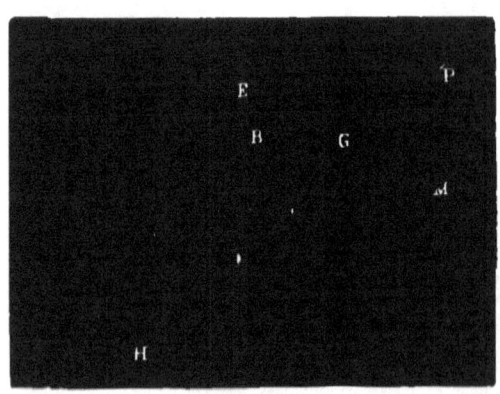

Fig. 36.

Draw tangents at A and A', meeting the tangent at P in G and H.

Then $\quad\quad A'T : TA = A'H : AG,\quad$ (Similar $\triangle^s$.)

and $\quad\quad A'M : MA = HP : PG$; (Parallel lines.)

but $\quad\quad A'H : AG = HP : PG$;

$$\quad\quad\quad\quad\quad\quad\quad\quad (Cor.\ 2,\ Prop.\ xxvi.)$$

$\therefore\ A'T : TA = A'M : MA$;

$\therefore\ A'T + TA : A'T - TA = A'M + MA : A'M - MA,$

or $\quad\quad 2CA : 2CT = 2CM : 2CA.$

hence $\quad\quad CM \times CT = CA^2.$

Cor. 1.—Conversely, if PM be an ordinate to any diameter AA', and CT be taken a third proportional to CM and CA; then PT will be a tangent to the hyperbola.

Cor. 2.—The tangents at the extremities of any double ordinate intersect on the diameter corresponding to that ordinate.

Cor. 3.—If a diameter be drawn through the intersection of two tangents to a hyperbola, it will bisect the chord of contact.

Cor. 4.—If the tangent at P meet a diameter of the conjugate hyperbola in D, and the ordinate PE be drawn; then CB is a mean proportional between CD and CE.

For $\qquad CT : CM = TD : DP \quad$ (Parallel lines.)

$\qquad\qquad\qquad\quad = DC : DE; \quad$ (Similar Δ^s.)

$\therefore CT \times CM : CM^2 = DC : DE$

or $\qquad CA^2 : CM^2 = DC : DE;$

$\therefore CA^2 : CM^2 - CA^2 = DC : DE - DC;$

$\therefore CB^2 : PM^2 \text{ or } CE^2 = DC : CE$

$\qquad\qquad\qquad\qquad$ (*Cor.* 2, Prop. XXII.)

$\qquad\qquad\qquad = DC \times CE : CE^2;$

hence $\qquad CB^2 = DC \times CE.$

Cor. 5.—Any diameter is cut harmonically by a tangent and the ordinate to the diameter drawn from the point of contact of the tangent.

Since $\qquad CM : CA = CA : CT.$

Comp. and Div.

$\qquad CM + CA : CM - CA = CA + CT : CA - CT,$

or $\qquad A'M : MA = A'T : TA.$

Cor. 6.—Any tangent to a hyperbola will be cut harmonically by two parallel tangents and the diameter passing through their points of contact.

For $\qquad MA' : MT : MA = PH : PT : PG.$

$\qquad\qquad\qquad\qquad\qquad$ (2, VI. Euclid.)

Proposition XXX.

If a variable tangent to a hyperbola meet two fixed parallel tangents, it will intercept segments on them whose rectangle is constant, and equal to the square of the parallel semi-diameter.

See Fig. 36.

Let PGH be the variable tangent meeting the fixed tangents AG, $A'H$ in the points G, H.

Draw the semi-diameter $CB \parallel AG$ or $A'H$; also draw PE an ordinate to BC.

Then $\qquad CT : CA = CA : CM.$ $\qquad$ (Prop. xxix.)

Division, $\qquad CT : TA = CA : AM.$

Alt. $\qquad CT : CA = TA : AM.$

Comp. $\qquad CT : TA' = TA : TM$;

$\therefore CD : A'H = AG : PM$; $\qquad$ (Similar $\triangle^s$.)

$\therefore A'H \times AG = CD \times PM$

$\qquad\qquad\qquad = CD \times CE$

$\qquad\qquad\qquad = CB^2.$ $\qquad$ (Cor. 4, Prop. xxix.)

Cor.—The rectangle under the segments of the variable tangent is equal to the square of the semi-diameter CQ drawn parallel to it.

For $\quad A'H : HP = CB : CQ = AG : PG$;
$\qquad\qquad\qquad\qquad\qquad$ (Cor. 2, Prop. xxvi.)

$\therefore A'H \times AG : HP \times PG = CB^2 : CQ^2.$

But $\quad A'H \times AG = CB^2$; $\therefore HP \times PG = CQ^2.$

Proposition XXXI.

The triangles CPT and CAK, formed by drawing tangents at the extremities of any two semi-diameters of a hyperbola, are equal in area.

Fig. 37.

Draw the ordinate PM.

Then $TC : CA = CA : CM$;   (Prop. xxix.)

$ = CK : CP$;   (Similar △'.)

$\therefore KT$ is $\parallel PA$;   (2, VI. Euclid.)

$\therefore \triangle PTA = \triangle PKA$;   (37, I. Euclid.)

hence $\triangle CPT = \triangle CAK$.

Cor.—If the ordinate AN be drawn from A to the semi-diameter CP produced; then

Area of $\triangle CMP$ = area of $\triangle CAN$.

For AN is $\parallel PT$; $\therefore TC : CA = PC : CN$;

hence $CA : CM = PC : CN$;

$\therefore MN$ is $\parallel PA$;   (2, VI. Euclid.)

$\therefore \triangle PAM = \triangle PAN$;   (31, I. Euclid.)

hence $\triangle CMP = \triangle CAN$.

Proposition XXXII.

If from the extremities of any two conjugate diameters CP, CQ, the ordinates PM, QN be drawn to *any* other diameter CA; then

$$CN^2 = AM \times MA';\text{ and }CM^2 = AM \times MA' + CA^2.$$

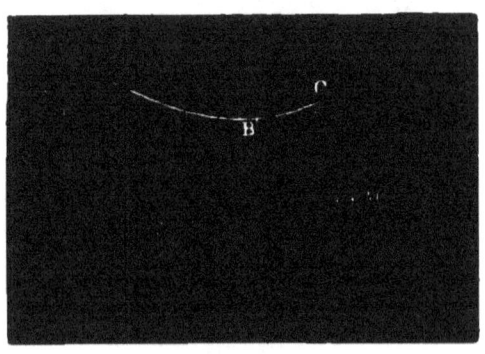

Fig. 38.

For $\quad PM^2 : CM^2 - CA^2 = CB^2 : CA^2$
$\hfill$ (*Cor.* 2, Prop. xxii.)

and $\quad QN^2 : CN + CA^2 = CB^2 : CA^2$;
$\hfill$ (*Cor.* 3, Prop. xxii.)

$\therefore PM^2 : QN^2 = CM^2 - CA^2 : CN^2 + CA^2$;

but $\quad PM : QN = TM : CN$; $\qquad$ (Similar $\triangle^s$.)

$\therefore TM^2 : CN^2 = CM^2 - CA^2 : CN^2 + CA^2$

$\qquad\qquad = CM^2 - CT \times CM : CN^2 + CA^2$
$\hfill$ (Prop. xxix.)

$\qquad\qquad = CM \times TM : CN^2 + CA^2$;
$\hfill$ (2, II. Euclid.)

$\therefore TM^2 : CM \times TM = CN^2 : CN^2 + CA^2$;

$\therefore TM : CM = CN^2 : CN^2 + CA^2$;

Div. $\quad TM : CT = CN^2 : CA^2$;

CHAP. III.] *The Hyperbola.* 161

$$\therefore TM \times CM : CT \times CM = CN^2 : CA^2;$$

but $\qquad CT \times CM = CA^2;$ \qquad (Prop. XXIX.)

$$\therefore TM \times CM = CN^2;$$

hence $\qquad CN^2 = CM^2 - CM \times CT$ \quad (2, II. Euclid.)

$$= CM^2 - CA^2$$

$$= AM \times MA'.$$

Also $\qquad CM^2 = CN^2 + CA^2$

$$= AM \times MA' + CA^2.$$

Cor. 1.— $\qquad CM^2 - CN^2 = CA^2.$

Cor. 2.— $\quad QN^2 - PM^2 = CB^2.$

For $\quad QN^2 : (CN^2 + CA^2)$ or $CM^2 = CB^2 : CA^2,$

and $\quad PM^2 : CM^2 - CA^2 = CB^2 : CA^2;$
$$\text{(Cor. 2, Prop. XXII.)}$$

$$\therefore QN^2 - PM^2 : CA^2 = CB^2 : CA^2;$$

hence $\qquad QN^2 - PM^2 = CB^2.$

Cor. 3.— $\quad CN : PM = CA : CB = CM : QN.$

For $CA^2 : CB^2 = CM^2 - CA^2 : PM^2 = CN^2 + CA^2 : QN^2,$

or $\quad CA^2 : CB^2 = CN^2 : PM^2 = CM^2 : QN^2;$ \qquad (*Cor.* 1.)

$$\therefore CA : CB = CN : PM = CM : QN.$$

Cor. 4.—The $\triangle\, CMP = \triangle\, CNQ$ in area.

For $\qquad CN : PM = CM : QN,$ \qquad (*Cor.* 3.)

and $\qquad \angle CMP = \angle CNQ;$

$$\therefore \triangle\, CMP = \triangle\, CNQ. \qquad \text{(15, VI. Euclid.)}$$

Proposition XXXIII.

If any tangent to a hyperbola meet any two conjugate diameters CP, CQ, the rectangle under its segments is equal to the square of the parallel semi-diameter CB.

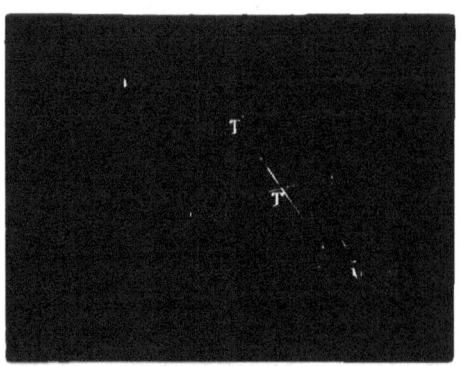

Fig. 39.

Draw the ordinates PM, QN, to the diameter passing through the point of contact.

Then $\quad CM : PM = CA : AT'$; (Similar $\triangle^s$.)

$\therefore CM \times AT' = PM \times CA$

$\qquad\qquad\quad = CB \times CN$;

$\qquad\qquad\qquad$ (Cor. 3, Prop. xxxii.)

$\therefore CM : CN = CB : AT'$.

Again, $\quad CN : NQ = CA : AT$. (Similar $\triangle^s$.)

$\therefore CN \times AT = CA \times NQ$

$\qquad\qquad\quad = CB \times CM$;

$\qquad\qquad\qquad$ (Cor. 3, Prop. xxxii.)

$\therefore CM : CN = AT : CB$;

hence $\quad CB : AT' = AT : CB$;

$\therefore AT \times AT' = CB^2$.

Proposition XXXIV.

Given in magnitude and position any two conjugate semi-diameters CA', CB', of a hyperbola, to find the axes.

Fig. 40.

Take $A'D$ a third proportional to CA' and CB'.

Bisect CD in H. Draw $HO \perp CD$, meeting $A'O$ drawn $\parallel CB'$ in O. With the centre O, and radius OC, describe a circle cutting OA' in T, T'; join CT, CT'; draw $A'M$, $A'N \perp CT$, CT', produced respectively. Take CA a mean proportional between CT and CM; also CB a mean proportional between CT' and CN. Then CA, CB are the axes.

For $\qquad CB'^2 = CA' \times A'D \qquad$ (Const.)

$\qquad\qquad\quad = TA' \times A'T'$; (36, III. Euclid.)

$\therefore$ CA and CB are conjugate diameters; (Prop. xxxiii.)
but $\angle TCT' = 90°$; $\therefore$ CA and CB are the axes in position.

Also $\quad CA^2 = CT \times CM$, and $CB^2 = CT' \times CN$;

$\therefore$ CA and CB represent the semiaxes in magnitude.

(Prop. ix.)

Proposition XXXV.

If any line *TS* be drawn parallel to the chord of contact of two tangents to a hyperbola, the segments *AT*, *BS*, intercepted between the curve and the tangents will be equal.

Fig. 41.

Draw the diameter of the parallel chords *BA*, *PQ*.
This diameter will pass through *O*.

(*Cor.* 2, Prop. xxix.)

Then, since $PM = MQ$; (*Cor.* 2, Prop. xxi.)

∴ $TN = NS$;

but $AN = NB$; (*Cor.* 2, Prop. xxi.)

hence $AT = BS$.

Cor. 1.—The tangent drawn parallel to the chord of contact of two other tangents is bisected at its point of contact.

Cor. 2.—If a line be drawn parallel to the chord of contact of two tangents, the segment intercepted between the tangents is bisected by the diameter passing through their intersection.

Proposition XXXVI.

Any line OA drawn through the intersection of two tangents to a hyperbola is cut harmonically by the curve and the chord of contact of the tangents.

Fig. 42.

Through A and A' draw TS, $T'S'\ \|\ PQ$, the chord of contact of the tangents. Then

$$TP^2 : T'P^2 = TA \times TB : T'A' \times T'B';$$
(*Cor.*, Prop. xxvi.)

but $\quad TB = AS$, and $T'B' = A'S'$; (Prop. xxxv.)

$$\therefore TP^2 : T'P^2 = TA \times AS : T'A' \times A'S';$$

but $\quad TA : T'A' = OA : OA' = AS : AS'$;
(Similar $\triangle^s$.)

$$\therefore TP^2 : T'P'^2 = OA^2 : OA'^2;$$

$$\therefore TP : T'P = OA : OA';$$

hence $\quad AO' : A'O = OA : OA'.$ (Parallel lines.)

Proposition XXXVII.

Any line drawn through the middle point of the chord of contact of two tangents to a hyperbola will be cut harmonically by the curve and the line drawn through the intersection of the tangents parallel to their chord of contact.

Fig. 43.

Let TP, TQ, be tangents, BB' any chord drawn through O', the middle point of PQ, and meeting the line drawn through $T \parallel PQ$ in O. Through B'' draw $B''B' \parallel$ to PQ; join TC.

Then TC produced will pass through O'.

(*Cor.* 3, Prop. xxix.)

$$B''M : B'B'' = \begin{Bmatrix} B''M : O'T' \\ O'T' : B'B'' \end{Bmatrix}$$

$$= \begin{Bmatrix} B''T : TT' \\ BT' : BB'' \end{Bmatrix} \quad \text{(Similar } \triangle\text{'s.)}$$

$$= B''T \times BT' : BB'' \times TT'.$$

But $TT' \times BB'' = 2B''T \times BT$;

∴ $B'B'' = 2B''M$;

∴ $B'M$ is an ordinate, and B' a point on the curve.

Now $TB : BT' = B''T : B''T'$; (Prop. xxxvii.)

∴ $OB : BO' = OB' : B'O'$. (Parallel lines.)

Proposition XXXVIII.

If any line cut the asymptotes of a hyperbola in Q, Q', and the curve in P, then the rectangle under PQ and PQ', is equal to the square of the semi-diameter CB' drawn parallel to QQ'.

Fig. 44.

Draw the semi-conjugate axis CB. Through Q draw $QRR' \parallel CB$.

Then $\quad QP \times QP' : QR \times QR' = CB'^2 : CB^2$;
$\hfill$ (*Cor.* 1, Prop. xxvi.)

but $\quad QP' = PQ'$, and $QR' = RQ''$; (Prop. xviii.)

$\therefore QP \times PQ' : QR \times RQ'' = CB'^2 : CB^2$;

but $\quad QR \times RQ'' = CB^2$; $\hfill$ (Prop. xvii.)

$\therefore QP \times PQ' = CB'^2$.

Cor.—The portion of any tangent intercepted between the asymptotes is = the parallel diameter.

For $CB'^2 = QP \times PQ' = DE^2$. (See *Cor.* 4, Prop. xviii., and Fig. 23.)

168 The Hyperbola. [CHAP. III.

Proposition XXXIX.

The parallelograms inscribed between the curve and its asymptotes are equal.

Fig. 45.

Let $CMPN$, $CM'P'N'$ be the parallelograms.
Join PP', and produce it to meet the asymptotes in Q, Q.

Then $\square\, CMPN : \square\, CM'P'N' = \begin{Bmatrix} PM : P'M' \\ PN : P'N' \end{Bmatrix}$

(23, VI. Euclid.)

$= \begin{Bmatrix} PQ : QP' \\ PQ' : P'Q' \end{Bmatrix}$.

(Similar $\triangle^s$.)

But $PQ = P'Q'$, and $PQ' = QP'$; (Prop. XVIII.)

hence $\square\, CMPN = \square\, CM'P'N'$.

Cor. 1.—The area of the sector of a hyperbola, made by joining any two points on the curve with the centre, is equal to the area included between the curve, one asymptote, and the lines drawn through the points parallel to the other asymptote.

For the $\triangle^s$ $CP'N'$ and CPN are equal in area, being halves of equal parallelograms; take them successively from the figure $CPP'N'$, and sector CPP' = figure $NPP'N'$.

Cor. 2.—The hyperbola continually approaches its asymptotes, but can never meet either at any finite distance.

For, the area of the $\square$ $CMPN$ being constant;

∴ PM varies inversely as PN or CM;

hence, as CM increases, PM will diminish.

And ∴ when CM becomes infinitely great, PM will be infinitely small.

Cor. 3.—The area of the $\triangle$ formed by any tangent to a hyperbola with the asymptotes is constant.

For the portion of a tangent at any point P intercepted between the asymptotes will be bisected at its point of contact (Prop. XVIII.);

∴ area of $\triangle$ RCS = 2 area of $\square$ $CMPN$, and ∴ constant.

Cor. 4.—If two hyperbolas have the same asymptotes, the segments of any line, drawn parallel to either asymptote, intercepted between the curves and the other asymptote, are in a constant ratio.

If the second hyperbola be supposed to cut PN in p, and $P'N'$ in p';

then $\square$ CP : $\square$ Cp = $\square$ CP' : $\square$ Cp';

∴ $PN : pN = P'N' : pN'$.

170   *The Hyperbola.*   [CHAP. III.

PROPOSITION XL.

If through any point P on a hyperbola lines OPO', QPQ' be drawn parallel to any two adjacent sides AD, DC, of an inscribed quadrilateral, meeting the opposite sides in O, O', and Q, Q': then

$PO \times PO' : PQ \times PQ'$ in a constant ratio.

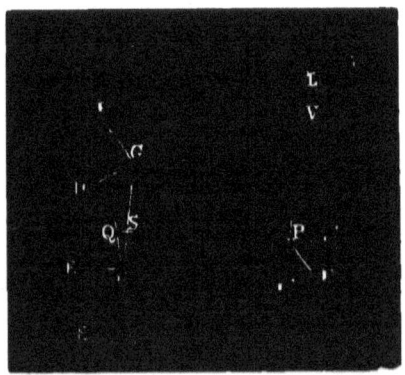

Fig. 46.

Through B and C draw BE and $CG \parallel AD$; join AG, and produce it to meet EB in H.

The diameter which bisects GC and BE will also bisect AD, and will ∴ bisect HK and LO'. Hence $BH = EK$, and $O'V = PL$.

Now , $OL : BH = LA : AH$

$= O'D : DK.$ (Parallel lines.)

Alt. $OL : O'D = BH : DK.$

Also PO' or $SC : SQ = BK : KC$;   (Similar △^{s}.)

∴ $OL \times PO' : O'D \times SQ = BH \times BK : DK \times KC$;

∴ $OL \times PO' : PQ' \times SQ = EK \times BK : DK \times KC$

$= O'V \times O'P : O'C \times O'D$

$= PL \times O'P : PS \times PQ'$;

[CHAP. III.] *The Hyperbola.* 171

$$\therefore OL \times PO' + PL \times O'P : PQ' \times SQ + PS \times PQ'$$
$$= EK \times BK : DK \times CK;$$

or $PO \times PO' : PQ \times PQ' = EK \times BK : DK \times KC;$

Now it is evident that the points A, B, C, D being fixed, E is also fixed, and $\therefore EK \times BK : DK \times KC$ in a constant ratio, and $\therefore PO \times PO' : PQ \times PQ'$ in a constant ratio.

Proposition XLI.

If from any point P on a hyperbola lines PR, PR', PS, PS' be drawn to the sides of an inscribed quadrilateral, making with them any constant angles; then the rectangles under the lines drawn to the opposite sides will be in a constant ratio.

Fig. 47.

Take any other point p on the curve.

Through the points P and p draw QPQ' and $qpq' \parallel DC$, and OPO', $poo' \parallel AD$; also $pr, pr', ps, ps' \parallel PR, PR', PS, PS'$ respectively.

Then $\qquad PR : pr = PQ : pq,\qquad$ (Similar $\triangle^s$.)

and $\qquad PR' : pr' = PQ' : pq'$;

$\therefore PR \times PR' : pr \times pr' = PQ \times PQ' : pq \times pq'$.

Similarly it may be proved that

$$PS \times PS' : ps \times ps' = PO \times PO' : po \times po';$$

but $\quad PQ \times PQ' : PO \times PO' = pq \times pq' : po \times po'$;

$\therefore PR \times PR' : PS \times PS' = pr \times pr' : ps \times ps'$.

Cor. 1.—The rectangle under the perpendiculars let fall from any point of a hyperbola on two opposite sides of an inscribed quadrilateral is in a constant ratio to the rectangle under the perpendiculars let fall on the other two sides.

Cor. 2.—If the points A and D coincide, also the points B and C, then the sides AD and BC become tangents, and the sides AB and CD coincide and become the chord of contact. Then the rectangle under the perpendiculars let fall from any point of an hyperbola on two fixed tangents is in a constant ratio to the square of the perpendicular let fall on their chord of contact.

Cor. 3.—If we suppose AB, CD to intersect in X, and AC, BD in Y; also that PR, PR', PS, PS', pr, pr', ps, ps' lie in the lines XY; then, regarding $ACBD$ as the quadrilateral, the above proportion becomes

$$PX^2 : PY^2 = pX^2 : pY^2;$$

$\therefore XY$ is cut harmonically by the curve.

Proposition XLII.

If two fixed tangents DP, DQ, to a hyperbola be cut by a diameter AB parallel to their chord of contact, and by a third variable tangent EF, the rectangle under the segments of the two fixed tangents intercepted between the diameter and the variable tangent is constant.

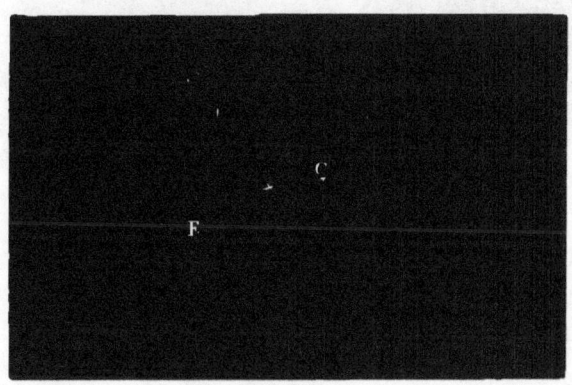

Fig. 48.

Join PC; produce it to meet the curve in R; join BR.

Since $\qquad CA = CB$; (Cor. 2, Prop. xxxv.)

∴ △'ACP and BCR are equal; (4, I. Euclid.)

∴ BR is = and ∥ AP, and ∴ a tangent.

If BR be supposed to be produced to meet FE produced in Z;
then $\qquad RB : PE = RZ : PD \qquad$ (Prop. xxx.)

$\qquad = RZ - RB : PD - PE = BZ : ED$;

∴ $AP : PE = BF : FD$;

∴ $AP : AP - PE = BF : BF - FD$;

∴ $AP : AE = BF : BD$;

∴ $AE \times BF = BD \times AP$, and ∴ constant.

Proposition XLIII.

To describe a conic section to pass through five given points.

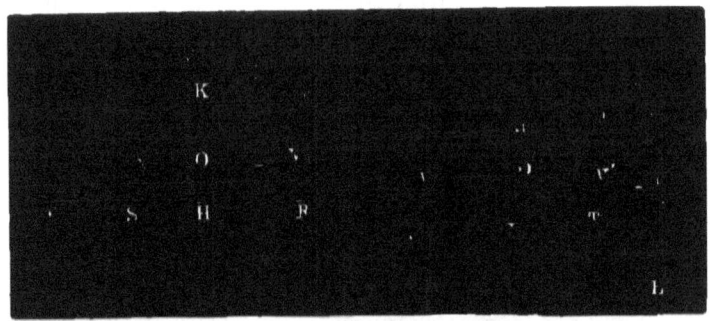

Fig. 49.   Fig. 50.

Join BD and CE. Through A draw $AQ \parallel BD$, and $AS \parallel CE$; in AQ find a point P, and in AS a point T, such that
$$BR \times RD : CR \times RE = AQ \times QP : CQ \times QE,$$
and $BR \times RD : CR \times RE = BS \times SD : AS \times ST$.

The points P and T will (Prop. XXII. Chap. II., and Cor. 1, Prop. XXVI. Chap. III.) be points on the curve.

If ∴ we bisect AT in N, and CE in M, and join MN, this line will evidently pass through the centre. Likewise bisect AP in K, and BD in H, and join KH; this line will also pass through the centre. Should MN and KH happen to be parallel the curve will be a parabola; if not, their point of intersection will determine the centre of the curve.

Now determine OV, such that
$$PK^2 : DH^2 = OV^2 - OK^2 : OV^2 - OH^2;$$
then (Cor. 2, Prop. XXII.) OV will be a semi-diameter to which PK and DH are ordinates; hence any number of ordinates and points on the curve may be found. Also the axes (by Prop. XXX. Chap. II., and Prop. XXXIV. Chap. III.).

CHAP. III.] *The Hyperbola.* 175

PROPOSITION XLIV.

To describe a conic section to touch five given right lines.

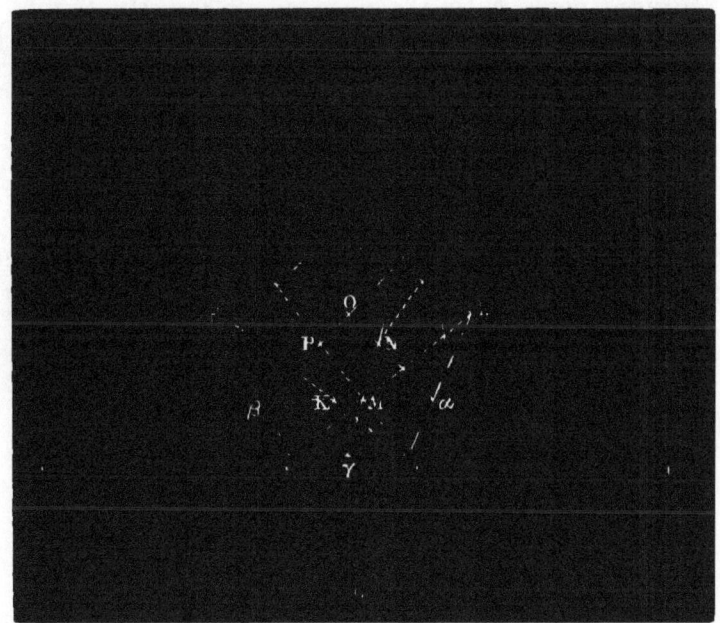

Fig. 51.

First, consider the quadrilateral *BFGH* formed by any four of the five tangents; draw the diagonals; let them intersect in *K*. Next consider the quadrilateral *AGHL* formed by leaving out another tangent; draw the diagonals; let them intersect in *M*.

In this way go round the figure, leaving out in succession each of the tangents, and three other points *N, O, P*, may be similarly found. The line joining *K, M*, produced will intersect the tangents in a, β; and the line joining *KP* produced will intersect the tangents in γ, δ, points on the curve.

176   *The Hyperbola.*   [CHAP. III.

For the intersection K of the diagonals of the quadrilateral $BFGH$ is (by Prop. XL.) a point on the chord of contact of the required conic with the lines FG, LH; also the intersection M of diagonals of the quadrilateral $AGHL$ is, by same Prop., a point on the chord of contact with the same lines; ∴ $a, \beta,$ are the points of contact of the required conic with these lines. Similarly it may be shown that $\gamma, \delta,$ and $\epsilon,$ are the points of contact with the other right lines.

PROPOSITION XLV.

If a right cone be cut by a plane which, when produced, cuts the opposite cone, the section will be a hyperbola.

Fig. 52.

Let the plane BVE drawn through the axis of the cone perpendicular to the plane of the section coincide with the plane of the paper, then both the section APM and the base BPE will be ⊥ the plane of the paper; ∴ the line

MP in which the section cuts the base is $\perp$ the plane of the paper, and $\therefore \perp BE$.

Hence $\qquad BM \times ME = MP^2.\qquad$ (35, III. Euclid.)

If now any other plane bpe be drawn $\parallel$ the base, meeting the section in pm, it can similarly be shown that mp is $\perp be$, and $\therefore bm \times me = mp^2$.

But $\qquad BM : bm = A'M : A'm;\qquad$ (Similar $\triangle^s$.)

also $\qquad ME : me = MA : mA;$

$\therefore BM \times ME : bm \times me = A'M \times MA : A'm \times mA,$

or $\qquad MP^2 : mp^2 = A'M \times MA : A'M \times mA.$

Hence the section is a hyperbola. $\qquad$ (Prop. x.)

Cor. 1.—The conjugate axis is a mean proportional between the diameters of the sections drawn through A and A' parallel to the base.

This follows, as in Chap. II., page 108.

Cor. 2.—The spheres inscribed in the cone, to touch the plane of the circle, will determine the foci.

This follows, as in Chap. II., page 109.

APPENDIX.

Many solutions having been given of Problem 9, page 110, I consider it advisable to add the following :—

Let PP' be the focal chord.

Draw the tangents, and also the normals at P and P'. Suppose the former to intersect in T, and the latter in N. Let the line drawn through $N \parallel$ to the axis intersect PP' in O, and FT in G.

Draw $NH \perp PP'$.

Then T is the centre exscribed to the $\triangle PF'P'$, and N is the centre of the inscribed circle;

$$\therefore F', N, F, \text{ are in directum};$$

also $\qquad TF \text{ is } \perp PP'; \qquad$ (Prop. XIII.)

$$\therefore P'F = PH.$$

(G. & H.'s Euclid, Appendix, Book IV.)

Now $\qquad P' . F'NPT$ is a harmonic pencil;

$$\therefore F . F'NPT \text{ is a harmonic pencil.}$$

$\therefore NG$, which is $\parallel F'F$ is bisected at O;

hence $\qquad FO = OH; \qquad$ (26, I. Euclid.)

$$\therefore P'O = OP.$$

In *Cors.* 2 and 3, Prop. XI., Chap. III., in order to describe the position of the point, it is assumed that the curve is concave towards the transverse axis. This property is fully demonstrated in *Cor.* 4, Prop. IX. It is to be noted that the essential property of the tangent, proved in Prop. III., does not depend on the assumption.

THE END.

APRIL 1879.

GENERAL LISTS OF NEW WORKS

PUBLISHED BY

Messrs. LONGMANS, GREEN & CO.

PATERNOSTER ROW, LONDON.

HISTORY, POLITICS, HISTORICAL MEMOIRS &c.

Armitage's Childhood of the English Nation. Fcp. 8vo. 2s. 6d.
Arnold's Lectures on Modern History. 8vo. 7s. 6d.
Bagehot's Literary Studies. 2 vols. 8vo. 28s.
Buckle's History of Civilisation. 3 vols. crown 8vo. 24s.
Chesney's Indian Polity. 8vo. 21s.
    — Waterloo Lectures. 8vo. 10s. 6d.
Digby's Famine Campaign in India. 2 vols. 8vo. 32s.
Durand's First Afghan War. Crown 8vo.
Epochs of Ancient History :—
         Beesly's Gracchi, Marius, and Sulla, 2s. 6d.
         Capes's Age of the Antonines, 2s. 6d.
         — Early Roman Empire, 2s. 6d.
         Cox's Athenian Empire, 2s. 6d.
         — Greeks and Persians, 2s. 6d.
         Curteis's Rise of the Macedonian Empire, 2s. 6d.
         Ihne's Rome to its Capture by the Gauls, 2s. 6d.
         Merivale's Roman Triumvirates, 2s. 6d.
         Sankey's Spartan and Theban Supremacies, 2s. 6d.
Epochs of English History :—
         Creighton's Shilling History of England (Introductory Volume). Fcp. 8vo. 1s.
         Browning's Modern England, 1820-1875, 9d.
         Cordery's Struggle against Absolute Monarchy, 1603-1688, 9d.
         Creighton's (Mrs.) England a Continental Power, 1066-1216, 9d.
         Creighton's (Rev. M.) Tudors and the Reformation, 1485-1603, 9d.
         Rowley's Rise of the People, 1215-1485, 9d.
         Rowley's Settlement of the Constitution, 1688-1778, 9d.
         Tancock's England during the American & European Wars, 1778-1820, 9d.
         York-Powell's Early England to the Conquest, 1s.
Epochs of Modern History :—
         Church's Beginning of the Middle Ages, 2s. 6d.
         Cox's Crusades, 2s. 6d.
         Creighton's Age of Elizabeth, 2s. 6d.
         Gairdner's Houses of Lancaster and York, 2s. 6d.
         Gardiner's Puritan Revolution, 2s. 6d.
         — Thirty Years' War, 2s. 6d.
         Hale's Fall of the Stuarts, 2s. 6d.
         Johnson's Normans in Europe, 2s. 6d.

London, LONGMANS & CO.

Epochs of Modern History—*continued*.
      Ludlow's War of American Independence, 2s. 6d.
      Morris's Age of Queen Anne, 2s. 6d.
      Seebohm's Protestant Revolution, 2s. 6d.
      Stubbs's Early Plantagenets, 2s. 6d.
      Warburton's Edward III., 2s. 6d.
Froude's English in Ireland in the 18th Century. 3 vols. 8vo. 48s.
 — History of England. 12 vols. 8vo. £8. 18s. 12 vols. crown 8vo. 72s.
 — Julius Cæsar, a Sketch. 8vo. 16s.
Gairdner's Richard III. and Perkin Warbeck. Crown 8vo. 10s. 6d.
Gardiner's England under Buckingham and Charles I., 1624–1628. 2 vols. 8vo. 24s.
 — Personal Government of Charles I., 1628–1637. 2 vols. 8vo. 24s.
Greville's Journal of the Reigns of George IV. & William IV. 3 vols. 8vo. 36s.
Hayward's Selected Essays. 2 vols. crown 8vo. 12s.
Hearn's Aryan Household. 8vo. 16s.
Howorth's History of the Mongols. Vol. I. Royal 8vo. 28s.
Ihne's History of Rome. 3 vols. 8vo. 45s.
Lecky's History of England. Vols. I. & II., 1700–1760. 8vo. 36s.
 — — — European Morals. 2 vols. crown 8vo. 16s.
 — Spirit of Rationalism in Europe. 2 vols. crown 8vo. 16s.
Lewes's History of Philosophy. 2 vols. 8vo. 32s.
Longman's Lectures on the History of England. 8vo. 15s.
 — Life and Times of Edward III. 2 vols. 8vo. 28s.
Macaulay's Complete Works. 8 vols. 8vo. £5. 5s.
 — History of England :—
    Student's Edition. 2 vols. cr. 8vo. 12s. | Cabinet Edition. 8 vols. post 8vo. 48s.
    People's Edition. 4 vols. cr. 8vo. 16s. | Library Edition. 5 vols. 8vo. £4.
Macaulay's Critical and Historical Essays. Cheap Edition. Crown 8vo. 3s. 6d.
    Cabinet Edition. 4 vols. post 8vo. 24s. | Library Edition. 3 vols. 8vo. 36s.
    People's Edition. 2 vols. cr. 8vo. 8s. | Student's Edition. 1 vol. cr. 8vo. 6s.
May's Constitutional History of England. 3 vols. crown 8vo. 18s.
 — Democracy in Europe. 2 vols. 8vo. 32s.
Merivale's Fall of the Roman Republic. 12mo. 7s. 6d.
 — General History of Rome, B.C. 753—A.D. 476. Crown 8vo. 7s. 6d.
 — History of the Romans under the Empire. 8 vols. post 8vo. 48s.
Phillips's Civil War in Wales and the Marches, 1642–1649. 8vo. 16s.
Prothero's Life of Simon de Montfort. Crown 8vo. 9s.
Rawlinson's Seventh Great Oriental Monarchy—The Sassanians. 8vo. 28s.
 — Sixth Oriental Monarchy—Parthia. 8vo. 16s.
Seebohm's Oxford Reformers—Colet, Erasmus, & More. 8vo. 14s.
Sewell's Popular History of France. Crown 8vo. 7s. 6d.
Short's History of the Church of England. Crown 8vo. 7s. 6d.
Smith's Carthage and the Carthaginians. Crown 8vo. 10s. 6d.
Taylor's Manual of the History of India. Crown 8vo. 7s. 6d.
Todd's Parliamentary Government in England. 2 vols. 8vo. 37s.
Trench's Realities of Irish Life. Crown 8vo. 2s. 6d.
Walpole's History of England. Vols. I. & II. 8vo. 36s.

BIOGRAPHICAL WORKS.

Burke's Vicissitudes of Families. 2 vols. crown 8vo. 21s.
Cates's Dictionary of General Biography. Medium 8vo. 25s.

Gleig's Life of the Duke of Wellington. Crown 8vo. 6s.
Jerrold's Life of Napoleon III. Vols. I. to III. 8vo. price 18s. each.
Jones's Life of Admiral Frobisher. Crown 8vo. 6s.
Lecky's Leaders of Public Opinion in Ireland. Crown 8vo. 7s. 6d.
Life (The) of Sir William Fairbairn. Crown 8vo. 18s.
Life (The) of Bishop Frampton. Crown 8vo. 10s. 6d.
Life (The) and Letters of Lord Macaulay. By his Nephew, G. Otto Trevelyan, M.P. Cabinet Edition, 2 vols. post 8vo. 12s. Library Edition, 2 vols. 8vo. 36s.
Marshman's Memoirs of Havelock. Crown 8vo. 3s. 6d.
Memoirs of Anna Jameson, by Gerardine Macpherson. 8vo. 12s. 6d.
Memorials of Charlotte Williams-Wynn. Crown 8vo. 10s. 6d.
Mendelssohn's Letters. Translated by Lady Wallace. 2 vols. cr. 8vo. 5s. each.
Mill's (John Stuart) Autobiography. 8vo. 7s. 6d.
Newman's Apologia pro Vita Sua. Crown 8vo. 6s.
Nohl's Life of Mozart. Translated by Lady Wallace. 2 vols. crown 8vo. 21s.
Pattison's Life of Casaubon. 8vo. 18s.
Spedding's Letters and Life of Francis Bacon. 7 vols. 8vo. £4. 4s.
Stephen's Essays in Ecclesiastical Biography. Crown 8vo. 7s. 6d.
Stigand's Life, Works &c. of Heinrich Heine. 2 vols. 8vo. 28s.
Zimmern's Life and Works of Lessing. Crown 8vo. 10s. 6d.

CRITICISM, PHILOSOPHY POLITY &c.

Amos's View of the Science of Jurisprudence. 8vo. 18s.
— Primer of the English Constitution. Crown 8vo. 6s.
Arnold's Manual of English Literature. Crown 8vo. 7s. 6d.
Bacon's Essays, with Annotations by Whately. 8vo. 10s. 6d.
— Works, edited by Spedding. 7 vols. 8vo. 73s. 6d.
Bain's Logic, Deductive and Inductive. Crown 8vo. 10s. 6d.
      PART I. Deduction, 4s. | PART II. Induction, 6s. 6d.
Blackley's German and English Dictionary. Post 8vo. 7s. 6d.
Bolland & Lang's Aristotle's Politics. Crown 8vo. 7s. 6d.
Bullinger's Lexicon and Concordance to the New Testament. Medium 8vo. 30s.
Comte's System of Positive Polity, or Treatise upon Sociology, translated:—
    VOL. I. General View of Positivism and its Introductory Principles. 8vo. 21s.
    VOL. II. Social Statics, or the Abstract Laws of Human Order. 14s.
    VOL. III. Social Dynamics, or General Laws of Human Progress. 21s.
    VOL. IV. Theory of the Future of Man; with Early Essays. 24s.
Congreve's Politics of Aristotle; Greek Text, English Notes. 8vo. 18s.
Contanseau's Practical French & English Dictionary. Post 8vo. 7s. 6d.
— Pocket French and English Dictionary. Square 18mo. 3s. 6d.
Dowell's Sketch of Taxes in England. VOL. I. to 1642. 8vo. 10s. 6d.
Farrar's Language and Languages. Crown 8vo. 6s.
Grant's Ethics of Aristotle, Greek Text, English Notes. 2 vols. 8vo. 32s.
Hodgson's Philosophy of Reflection. 2 vols. 8vo. 21s.
Kalisch's Historical and Critical Commentary on the Old Testament; with a New Translation. Vol. I. *Genesis*, 8vo. 18s. or adapted for the General Reader, 12s. Vol. II. *Exodus*, 15s. or adapted for the General Reader, 12s. Vol. III. *Leviticus*, Part I. 15s. or adapted for the General Reader, 8s. Vol. IV. *Leviticus*, Part II. 15s. or adapted for the General Reader, 8s.

London, LONGMANS & CO.

Latham's Handbook of the English Language. Crown 8vo. 6s.
— English Dictionary. 1 vol. medium 8vo. 24s. 4 vols. 4to. £7.
Lewis on Authority in Matters of Opinion. 8vo. 14s.
Liddell & Scott's Greek-English Lexicon. Crown 4to. 36s.
— — — Abridged Greek-English Lexicon. Square 12mo. 7s. 6d.
Longman's Pocket German and English Dictionary. 18mo. 5s.
Macaulay's Speeches corrected by Himself. Crown 8vo. 3s. 6d.
Macleod's Economical Philosophy. Vol. I. 8vo. 15s. Vol. II. Part I. 12s.
Mill on Representative Government. Crown 8vo. 2s.
— — Liberty. Post 8vo. 7s. 6d. Crown 8vo. 1s. 4d.
Mill's Dissertations and Discussions. 4 vols. 8vo. 46s. 6d.
— Essays on Unsettled Questions of Political Economy. 8vo. 6s. 6d.
— Examination of Hamilton's Philosophy. 8vo. 16s.
— Logic, Ratiocinative and Inductive. 2 vols. 8vo. 25s.
— Phenomena of the Human Mind. 2 vols. 8vo. 28s.
— Principles of Political Economy. 2 vols. 8vo. 30s. 1 vol. cr. 8vo. 5s.
— Subjection of Women. Crown 8vo. 6s.
— Utilitarianism. 8vo. 5s.
Morell's Philosophical Fragments. Crown 8vo. 5s.
Müller's (Max) Lectures on the Science of Language. 2 vols. crown 8vo. 16s.
— Hibbert Lectures on the Origin and Growth of Religion. 8vo. 10s. 6d.
Noiré on Max Müller's Philosophy of Language. 8vo. 6s.
Rich's Dictionary of Roman and Greek Antiquities. Crown 8vo. 7s. 6d.
Roget's Thesaurus of English Words and Phrases. Crown 8vo. 10s. 6d.
Sandars's Institutes of Justinian, with English Notes. 8vo. 18s.
Swinbourne's Picture Logic. Post 8vo. 5s.
Thomson's Outline of Necessary Laws of Thought. Crown 8vo. 6s.
Tocqueville's Democracy in America, translated by Reeve. 2 vols. crown 8vo. 16s.
Twiss's Law of Nations, 8vo. in Time of Peace, 12s. in Time of War, 21s.
Whately's Elements of Logic. 8vo. 10s. 6d. Crown 8vo. 4s. 6d.
— — — Rhetoric. 8vo. 10s. 6d. Crown 8vo. 4s. 6d.
— English Synonymes. Fcp. 8vo. 3s.
White & Riddle's Large Latin-English Dictionary. 4to. 28s.
White's College Latin-English Dictionary. Medium 8vo. 15s.
— Junior Student's Complete Latin-English and English-Latin Dictionary. Square 12mo. 12s.
Separately { The English-Latin Dictionary, 5s. 6d.
{ The Latin-English Dictionary, 7s. 6d.
White's Middle-Class Latin-English Dictionary. Fcp. 8vo. 8s.
Williams's Nicomachean Ethics of Aristotle translated. Crown 8vo. 7s. 6d
Yonge's Abridged English-Greek Lexicon. Square 12mo. 8s. 6d.
— Large English-Greek Lexicon. 4to. 21s.
Zeller's Socrates and the Socratic Schools. Crown 8vo. 10s. 6d.
— Stoics, Epicureans, and Sceptics. Crown 8vo. 14s.
— Plato and the Older Academy. Crown 8vo. 18s.

MISCELLANEOUS WORKS & POPULAR METAPHYSICS.

Arnold's (Dr. Thomas) Miscellaneous Works. 8vo. 7s. 6d.
Bain's Emotions and the Will. 8vo. 15s.

General Lists of New Works. 5

Bain's Mental and Moral Science. Crown 8vo. 10s. 6d. Or separately: Part I.
Mental Science, 6s. 6d. Part II. Moral Science, 4s. 6d.
— Senses and the Intellect. 8vo. 15s.
Buckle's Miscellaneous and Posthumous Works. 3 vols. 8vo. 52s. 6d.
Conington's Miscellaneous Writings. 2 vols. 8vo. 28s.
Edwards's Specimens of English Prose. 16mo. 2s. 6d.
Froude's Short Studies on Great Subjects. 3 vols. crown 8vo. 18s.
German Home Life, reprinted from *Fraser's Magazine*. Crown 8vo. 6s.
Hume's Essays, edited by Green & Grose. 2 vols. 8vo. 28s.
— Treatise of Human Nature, edited by Green & Grose. 2 vols. 8vo. 28s.
Macaulay's Miscellaneous Writings. 2 vols. 8vo. 21s. 1 vol. crown 8vo. 4s. 6d.
— Writings and Speeches. Crown 8vo. 6s.
Mill's Analysis of the Phenomena of the Human Mind. 2 vols. 8vo. 28s.
Müller's (Max) Chips from a German Workshop. 4 vols. 8vo. 58s.
Mullinger's Schools of Charles the Great. 8vo. 7s. 6d.
Rogers's Defence of the Eclipse of Faith Fcp. 8vo. 3s. 6d.
— Eclipse of Faith. Fcp. 8vo. 5s.
Selections from the Writings of Lord Macaulay. Crown 8vo. 6s.
The Essays and Contributions of A. K. H. B. Crown 8vo.

   Autumn Holidays of a Country Parson. 3s. 6d.
   Changed Aspects of Unchanged Truths. 3s. 6d.
   Common-place Philosopher in Town and Country. 3s. 6d.
   Counsel and Comfort spoken from a City Pulpit. 3s. 6d.
   Critical Essays of a Country Parson. 3s. 6d.
   Graver Thoughts of a Country Parson. Three Series, 3s. 6d. each.
   Landscapes, Churches, and Moralities. 3s. 6d.
   Leisure Hours in Town. 3s. 6d.
   Lessons of Middle Age. 3s. 6d.
   Present-day Thoughts. 3s. 6d.
   Recreations of a Country Parson. Three Series, 3s. 6d. each.
   Seaside Musings on Sundays and Week-Days. 3s. 6d.
   Sunday Afternoons in the Parish Church of a University City. 3s. 6d.

Wit and Wisdom of the Rev. Sydney Smith 16mo. 3s. 6d.

ASTRONOMY, METEOROLOGY, POPULAR GEOGRAPHY &c.

Dove's Law of Storms, translated by Scott. 8vo. 10s. 6d.
Herschel's Outlines of Astronomy. Square crown 8vo. 12s.
Keith Johnston's Dictionary of Geography, or Gazetteer. 8vo. 42s.
Nelson's Work on the Moon. Medium 8vo. 31s. 6d.
Proctor's Essays on Astronomy. 8vo. 12s.
— Larger Star Atlas. Folio, 15s. or Maps only, 12s. 6d.
— Moon. Crown 8vo. 10s. 6d.
— New Star Atlas. Crown 8vo. 5s.
— Orbs Around Us. Crown 8vo. 7s. 6d.
— Other Worlds than Ours. Crown 8vo. 10s. 6d.
— Saturn and its System. 8vo. 14s.
— Sun. Crown 8vo. 14s.
— Transits of Venus, Past and Coming. Crown 8vo. 8s. 6d.
— Treatise on the Cycloid and Cycloidal Curves. Crown 8vo. 10s. 6d.

London, LONGMANS & CO.

Proctor's Universe of Stars. 8vo. 10s. 6d.
Schellen's Spectrum Analysis. 8vo. 28s.
Smith's Air and Rain. 8vo. 24s.
The Public Schools Atlas of Ancient Geography. Imperial 8vo. 7s. 6d.
— — — Atlas of Modern Geography. Imperial 8vo. 5s.
Webb's Celestial Objects for Common Telescopes. New Edition in preparation.

NATURAL HISTORY & POPULAR SCIENCE.

Arnott's Elements of Physics or Natural Philosophy. Crown 8vo. 12s. 6d.
Brande's Dictionary of Science, Literature, and Art. 3 vols. medium 8vo. 63s.
Decaisne and Le Maout's General System of Botany. Imperial 8vo. 31s. 6d.
Evans's Ancient Stone Implements of Great Britain. 8vo. 28s.
Ganot's Elementary Treatise on Physics, by Atkinson. Large crown 8vo. 15s.
— Natural Philosophy, by Atkinson. Crown 8vo. 7s. 6d.
Gore's Art of Scientific Discovery. Crown 8vo. 15s.
Grove's Correlation of Physical Forces. 8vo. 15s.
Hartwig's Aerial World. 8vo. 10s. 6d.
— Polar World. 8vo. 10s. 6d.
— Sea and its Living Wonders. 8vo. 10s. 6d.
— Subterranean World. 8vo. 10s. 6d.
— Tropical World. 8vo. 10s. 6d.
Haughton's Principles of Animal Mechanics. 8vo. 21s.
Heer's Primæval World of Switzerland. 2 vols. 8vo. 16s.
Helmholtz's Lectures on Scientific Subjects. 8vo. 12s. 6d.
Helmholtz on the Sensations of Tone, by Ellis. 8vo. 36s.
Hemsley's Handbook of Trees, Shrubs, & Herbaceous Plants. Medium 8vo. 12s.
Hullah's Lectures on the History of Modern Music. 8vo. 8s. 6d.
— Transition Period of Musical History. 8vo. 10s. 6d.
Keller's Lake Dwellings of Switzerland, by Lee. 2 vols. royal 8vo. 42s.
Kirby and Spence's Introduction to Entomology. Crown 8vo. 5s.
Lloyd's Treatise on Magnetism. 8vo. 10s. 6d.
— — on the Wave-Theory of Light. 8vo. 10s. 6d.
London's Encyclopædia of Plants. 8vo. 42s.
Lubbock on the Origin of Civilisation & Primitive Condition of Man. 8vo. 18s.
Macalister's Zoology and Morphology of Vertebrate Animals. 8vo. 10s. 6d.
Nicols' Puzzle of Life. Crown 8vo. 3s. 6d.
Owen's Comparative Anatomy and Physiology of the Vertebrate Animals. 3 vols. 8vo. 73s. 6d.
Proctor's Light Science for Leisure Hours. 2 vols. crown 8vo. 7s. 6d. each.
Rivers's Rose Amateur's Guide. Fcp. 8vo. 4s. 6d.
Stanley's Familiar History of Birds. Fcp. 8vo. 3s. 6d.
Text-Books of Science, Mechanical and Physical.
    Abney's Photography, small 8vo. 3s. 6d.
    Anderson's (Sir John) Strength of Materials, 3s. 6d.
    Armstrong's Organic Chemistry, 3s. 6d.
    Barry's Railway Appliances, 3s. 6d.
    Bloxam's Metals, 3s. 6d.
    Goodeve's Elements of Mechanism, 3s. 6d.
    — Principles of Mechanics, 3s. 6d.
    Gore's Electro-Metallurgy, 6s.
    Griffin's Algebra and Trigonometry, 3s. 6d.

General Lists of New Works.

Text-Books of Science—*continued.*

 Jenkin's Electricity and Magnetism, 3s. 6d.
 Maxwell's Theory of Heat, 3s. 6d.
 Merrifield's Technical Arithmetic and Mensuration, 3s. 6d.
 Miller's Inorganic Chemistry, 3s. 6d.
 Preece & Sivewright's Telegraphy, 3s. 6d.
 Rutley's Study of Rocks, 4s. 6d.
 Shelley's Workshop Appliances, 3s. 6d.
 Thomé's Structural and Physiological Botany, 6s.
 Thorpe's Quantitative Chemical Analysis, 4s. 6d.
 Thorpe & Muir's Qualitative Analysis, 3s. 6d.
 Tilden's Chemical Philosophy, 3s. 6d.
 Unwin's Machine Design, 3s. 6d.
 Watson's Plane and Solid Geometry, 3s. 6d.

Tyndall on Sound. Crown 8vo. 10s. 6d.
— Contributions to Molecular Physics. 8vo. 16s.
— Fragments of Science. New Edit. 2 vols. crown 8vo. [*In the press.*
— Heat a Mode of Motion. Crown 8vo.
— Lectures on Electrical Phenomena. Crown 8vo. 1s. sewed, 1s. 6d. cloth.
— Lectures on Light. Crown 8vo. 1s. sewed, 1s. 6d. cloth.
— Lectures on Light delivered in America. Crown 8vo. 7s. 6d.
— Lessons in Electricity. Crown 8vo. 2s. 6d.

Von Cotta on Rocks, by Lawrence. Post 8vo. 14s.
Woodward's Geology of England and Wales. Crown 8vo. 14s.
Wood's Bible Animals. With 112 Vignettes. 8vo. 14s.
— Homes Without Hands. 8vo. 14s.
— Insects Abroad. 8vo. 14s.
— Insects at Home. With 700 Illustrations. 8vo. 14s.
— Out of Doors, or Articles on Natural History. Crown 8vo. 7s. 6d.
— Strange Dwellings. With 60 Woodcuts. Crown 8vo. 7s. 6d.

CHEMISTRY & PHYSIOLOGY.

Auerbach's Anthracen, translated by W. Crookes, F.R.S. 8vo. 12s.
Buckton's Health in the House; Lectures on Elementary Physiology. Fcp. 8vo. 2s.
Crookes's Handbook of Dyeing and Calico Printing. 8vo. 42s.
— Select Methods in Chemical Analysis. Crown 8vo. 12s. 6d.
Kingzett's Animal Chemistry. 8vo. 18s.
— History, Products and Processes of the Alkali Trade. 8vo. 12s.
Miller's Elements of Chemistry, Theoretical and Practical. 3 vols. 8vo. Part I. Chemical Physics, 16s. Part II. Inorganic Chemistry, 24s. Part III. Organic Chemistry, New Edition in the press.
Watts's Dictionary of Chemistry. 7 vols. medium 8vo. £10. 16s. 6d.
— Third Supplementary Volume, in Two Parts. PART I. 36s.

THE FINE ARTS & ILLUSTRATED EDITIONS.

Bewick's Select Fables of Æsop and others. Crown 8vo. 7s. 6d. demy 8vo. 18s.
Doyle's Fairyland; Pictures from the Elf-World. Folio, 15s.
Jameson's Sacred and Legendary Art. 6 vols. square crown 8vo.
 Legends of the Madonna. 1 vol. 21s.
 — — — Monastic Orders. 1 vol. 21s.
 — — — Saints and Martyrs. 2 vols. 31s. 6d.
 — — — Saviour. Completed by Lady Eastlake. 2 vols. 42s.

London, LONGMANS & CO.

Longman's Three Cathedrals Dedicated to St. Paul. Square crown 8vo. 21s.
Macaulay's Lays of Ancient Rome. With 90 Illustrations. Fcp. 4to. 21s.
Macfarren's Lectures on Harmony. 8vo. 12s.
Miniature Edition of Macaulay's Lays of Ancient Rome. Imp. 16mo. 10s. 6d.
Moore's Irish Melodies. With 161 Plates by D. Maclise, R.A. Super-royal 8vo. 21s.
   — Lalla Rookh. Tenniel's Edition. With 68 Illustrations. Fcp. 4to. 21s.
Northcote and Brownlow's Roma Sotterranea. PART I. 8vo. 24s.
Perry on Greek and Roman Sculpture. 8vo. [*In preparation.*
Redgrave's Dictionary of Artists of the English School. 8vo. 16s.

THE USEFUL ARTS, MANUFACTURES &c.

Bourne's Catechism of the Steam Engine. Fcp. 8vo. 6s.
   — Examples of Steam, Air, and Gas Engines. 4to. 70s.
   — Handbook of the Steam Engine. Fcp. 8vo. 9s.
   — Recent Improvements in the Steam Engine. Fcp. 8vo. 6s.
   — Treatise on the Steam Engine. 4to. 42s.
Cresy's Encyclopædia of Civil Engineering. 8vo. 42s.
Culley's Handbook of Practical Telegraphy. 8vo. 16s.
Eastlake's Household Taste in Furniture, &c. Square crown 8vo. 14s.
Fairbairn's Useful Information for Engineers. 3 vols. crown 8vo. 31s. 6d.
   — Applications of Cast and Wrought Iron. 8vo. 16s.
   — Mills and Millwork. 1 vol. 8vo. 25s.
Gwilt's Encyclopædia of Architecture. 8vo. 52s. 6d.
Hobson's Amateur Mechanics Practical Handbook. Crown 8vo. 2s. 6d.
Hoskold's Engineer's Valuing Assistant. 8vo. 31s. 6d.
Kerl's Metallurgy, adapted by Crookes and Röhrig. 3 vols. 8vo. £4. 19s.
Loudon's Encyclopædia of Agriculture. 8vo. 21s.
   — — — Gardening. 8vo. 21s.
Mitchell's Manual of Practical Assaying. 8vo. 31s. 6d.
Northcott's Lathes and Turning. 8vo. 18s.
Payen's Industrial Chemistry, translated from Stohmann and Engler's German Edition, by Dr. J. D. Barry. Edited by B. H. Paul, Ph.D. 8vo. 42s.
Stoney's Theory of Strains in Girders. Roy. 8vo. 36s.
Thomas on Coal, Mine-Gases and Ventilation. Crown 8vo. 10s. 6d.
Ure's Dictionary of Arts, Manufactures, & Mines. 4 vols. medium 8vo. £7. 7s.

RELIGIOUS & MORAL WORKS.

Abbey & Overton's English Church in the Eighteenth Century. 2 vols. 8vo. 36s.
Arnold's (Rev. Dr. Thomas) Sermons. 6 vols. crown 8vo. 5s. each.
Bishop Jeremy Taylor's Entire Works. With Life by Bishop Heber. Edited by the Rev. C. P. Eden. 10 vols. 8vo. £5. 5s.
Boultbee's Commentary on the 39 Articles. Crown 8vo. 6s.
Browne's (Bishop) Exposition of the 39 Articles. 8vo. 16s.
Conybeare & Howson's Life and Letters of St. Paul :—
   Library Edition, with all the Original Illustrations, Maps, Landscapes on Steel, Woodcuts, &c. 2 vols. 4to. 42s.
   Intermediate Edition, with a Selection of Maps, Plates, and Woodcuts. 2 vols. square crown 8vo. 21s.
   Student's Edition, revised and condensed, with 46 Illustrations and Maps. 1 vol. crown 8vo. 9s.
Colenso's Lectures on the Pentateuch and the Moabite Stone. 8vo. 12s.

London, LONGMANS & CO.

General Lists of New Works.

Colenso on the Pentateuch and Book of Joshua. Crown 8vo. 6s.
— — PART VII. completion of the larger Work. 8vo. 24s.
D'Aubigné's Reformation in Europe in the Time of Calvin. 8 vols. 8vo. £6. 12s.
Drummond's Jewish Messiah. 8vo. 15s.
Ellicott's (Bishop) Commentary on St. Paul's Epistles. 8vo. Galatians, 8s. 6d. Ephesians, 8s. 6d. Pastoral Epistles, 10s. 6d. Philippians, Colossians, and Philemon, 10s. 6d. Thessalonians, 7s. 6d.
Ellicott's Lectures on the Life of our Lord. 8vo. 12s.
Ewald's History of Israel, translated by Carpenter. 5 vols. 8vo. 63s.
— Antiquities of Israel, translated by Solly. 8vo. 12s. 6d.
Goldziher's Mythology among the Hebrews. 8vo. 16s.
Jukes's Types of Genesis. Crown 8vo. 7s. 6d.
— Second Death and the Restitution of all Things. Crown 8vo. 3s. 6d.
Kalisch's Bible Studies. PART I. the Prophecies of Balaam. 8vo. 10s. 6d.
— — — PART II. the Book of Jonah. 8vo. 10s. 6d.
Keith's Evidence of the Truth of the Christian Religion derived from the Fulfilment of Prophecy. Square 8vo. 12s. 6d. Post 8vo. 6s.
Kuenen on the Prophets and Prophecy in Israel. 8vo. 21s.
Lyra Germanica. Hymns translated by Miss Winkworth. Fcp. 8vo. 5s.
Manning's Temporal Mission of the Holy Ghost. 8vo. 8s. 6d.
Martineau's Endeavours after the Christian Life. Crown 8vo. 7s. 6d.
— Hymns of Praise and Prayer. Crown 8vo. 4s. 6d. 32mo. 1s. 6d.
— Sermons; Hours of Thought on Sacred Things. Crown 8vo. 7s. 6d.
Merivale's (Dean) Lectures on Early Church History. Crown 8vo.
Mill's Three Essays on Religion. 8vo. 10s. 6d.
Monsell's Spiritual Songs for Sundays and Holidays. Fcp. 8vo. 5s. 18mo. 2s.
Müller's (Max) Lectures on the Science of Religion. Crown 8vo. 10s. 6d.
Newman's Apologia pro Vita Sua. Crown 8vo. 6s.
O'Conor's New Testament Commentaries. Crown 8vo. Epistle to the Romans. 3s. 6d. Epistle to the Hebrews, 4s. 6d. St. John's Gospel, 10s. 6d.
One Hundred Holy Songs, &c. Square fcp. 8vo. 2s. 6d.
Passing Thoughts on Religion. By Miss Sewell. Fcp. 8vo. 3s. 6d.
Sewell's (Miss) Preparation for the Holy Communion. 32mo. 3s.
Shipley's Ritual of the Altar. Imperial 8vo. 42s.
Supernatural Religion. 3 vols. 8vo. 38s.
Thoughts for the Age. By Miss Sewell. Fcp. 8vo. 3s. 6d.
Vaughan's Trident, Crescent, and Cross; the Religious History of India. 8vo. 9s. 6d.
Whately's Lessons on the Christian Evidences. 18mo. 6d.
White's Four Gospels in Greek, with Greek-English Lexicon. 32mo. 5s.

TRAVELS, VOYAGES &c.

Ball's Alpine Guide. 3 vols. post 8vo. with Maps and Illustrations:—I. Western Alps, 6s. 6d. II. Central Alps, 7s. 6d. III. Eastern Alps, 10s. 6d.
Ball on Alpine Travelling, and on the Geology of the Alps, 1s.
Baker's Rifle and the Hound in Ceylon. Crown 8vo. 7s. 6d.
— Eight Years in Ceylon. Crown 8vo. 7s. 6d.
Bent's Freak of Freedom, or the Republic of San Marino. Crown 8vo.
Brassey's Voyage in the Yacht 'Sunbeam.' Crown 8vo. 7s. 6d. 8vo. 21s.
Edwards's (A. B.) Thousand Miles up the Nile. Imperial 8vo. 42s.

London, LONGMANS & CO.

Evans's Illyrian Letters. Post 8vo. 7s. 6d.
Grohman's Tyrol and the Tyrolese. Crown 8vo. 6s.
Indian Alps (The). By a Lady Pioneer. Imperial 8vo. 42s.
Lefroy's Discovery and Early Settlement of the Bermuda Islands. 2 vols. royal 8vo. 60s.
Miller and Skertchley's Fenland Past and Present. Royal 8vo. 31s. 6d. Large Paper, 50s.
Noble's Cape and South Africa. Fcp. 8vo. 3s. 6d.
Packe's Guide to the Pyrenees, for Mountaineers. Crown 8vo. 7s. 6d.
The Alpine Club Map of Switzerland. In four sheets. 42s.
Wood's Discoveries at Ephesus. Imperial 8vo. 63s.

WORKS OF FICTION.

Becker's Charicles; Private Life among the Ancient Greeks. Post 8vo. 7s. 6d.
— Gallus; Roman Scenes of the Time of Augustus. Post 8vo. 7s. 6d.

Cabinet Edition of Stories and Tales by Miss Sewell:—

Amy Herbert, 2s. 6d.
Cleve Hall, 2s. 6d.
The Earl's Daughter, 2s. 6d.
Experience of Life, 2s. 6d.
Gertrude, 2s. 6d.
Ivors, 2s. 6d.
Katharine Ashton, 2s. 6d.
Laneton Parsonage, 3s. 6d.
Margaret Percival, 3s. 6d.
Ursula, 3s. 6d.

Novels and Tales by the Right Hon. the Earl of Beaconsfield, K.G. Cabinet Edition, complete in Ten Volumes, crown 8vo. price £3.

Lothair, 6s.
Coningsby, 6s.
Sybil, 6s.
Tancred, 6s.
Venetia, 6s.
Henrietta Temple, 6s.
Contarini Fleming, 6s.
Alroy, Ixion, &c. 6s.
The Young Duke, &c. 6s.
Vivian Grey, 6s.

The Modern Novelist's Library. Each Work in crown 8vo. A Single Volume, complete in itself, price 2s. boards, or 2s. 6d. cloth:—

By the Earl of Beaconsfield, K.G.
Lothair.
Coningsby.
Sybil.
Tancred.
Venetia.
Henrietta Temple.
Contarini Fleming.
Alroy, Ixion, &c.
The Young Duke, &c.
Vivian Grey.
By Anthony Trollope.
Barchester Towers.
The Warden.
By the Author of 'the Rose Garden.'
Unawares.

By Major Whyte-Melville.
Digby Grand.
General Bounce.
Kate Coventry.
The Gladiators.
Good for Nothing.
Holmby House.
The Interpreter.
The Queen's Maries.
By the Author of 'the Atelier du Lys.'
Mademoiselle Mori.
The Atelier du Lys.
By Various Writers.
Atherstone Priory.
The Burgomaster's Family.
Elsa and her Vulture.
The Six Sisters of the Valley.

Lord Beaconsfield's Novels and Tales. 10 vols. cloth extra, gilt edges, 30s.

Whispers from Fairy Land. By the Right Hon. E. H. Knatchbull-Hugessen M.P. With Nine Illustrations. Crown 8vo. 3s. 6d.

Higgledy-Piggledy; or, Stories for Everybody and Everybody's Children. By the Right Hon. E. M. Knatchbull-Hugessen, M.P. With Nine Illustrations from Designs by R. Doyle. Crown 8vo. 3s. 6d.

London, LONGMANS & CO.

POETRY & THE DRAMA.

Bailey's Festus, a Poem. Crown 8vo. 12s. 6d.
Bowdler's Family Shakspeare. Medium 8vo. 14s. 6 vols. fcp. 8vo. 21s.
Brian Boru, a Tragedy, by J. T. B. Crown 8vo. 6s.
Cayley's Iliad of Homer, Homometrically translated. 8vo. 12s. 6d.
Conington's Æneid of Virgil, translated into English Verse. Crown 8vo. 9s.
Cooper's Tales from Euripides. Small 8vo.
Edwards's Poetry-Book of Elder Poets. 16mo. 2s. 6d.
— Poetry-Book of Modern Poets. 16mo. 2s. 6d.
Ingelow's Poems. First Series. Illustrated Edition. Fcp. 4to. 21s.
Macaulay's Lays of Ancient Rome, with Ivry and the Armada. 16mo. 3s. 6d.
Petrarch's Sonnets and Stanzas, translated by C. B. Cayley, B.A. Crown 8vo. 10s. 6d.
Poems. By Jean Ingelow. 2 vols. fcp. 8vo. 10s.
    First Series. 'Divided,' 'The Star's Monument,' &c. 5s.
    Second Series. 'A Story of Doom,' 'Gladys and her Island,' &c. 5s.
Southey's Poetical Works. Medium 8vo. 14s.
Yonge's Horatii Opera, Library Edition. 8vo. 21s.

RURAL SPORTS, HORSE & CATTLE MANAGEMENT &c.

Blaine's Encyclopædia of Rural Sports. 8vo. 21s.
Dobson on the Ox, his Diseases and their Treatment. Crown 8vo. 7s. 6d.
Fitzwygram's Horses and Stables. 8vo. 10s. 6d.
Francis's Book on Angling, or Treatise on Fishing. Post 8vo. 15s.
Malet's Annals of the Road, and Nimrod's Essays on the Road. Medium 8vo. 21s.
Miles's Horse's Foot, and How to Keep it Sound. Imperial 8vo. 12s. 6d.
— Plain Treatise on Horse-Shoeing. Post 8vo. 2s. 6d.
— Stables and Stable-Fittings. Imperial 8vo. 15s.
— Remarks on Horses' Teeth. Post 8vo. 1s. 6d.
Nevile's Horses and Riding. Crown 8vo. 6s.
Reynardson's Down the Road. Medium 8vo. 21s.
Ronalds's Fly-Fisher's Entomology. 8vo. 14s.
Stonehenge's Dog in Health and Disease. Square crown 8vo. 7s. 6d.
— Greyhound. Square crown 8vo. 15s.
Youatt's Work on the Dog. 8vo. 12s. 6d.
— — — — Horse. 8vo. 6s.
Wilcocks's Sea-Fisherman. Post 8vo. 12s. 6d.

WORKS OF UTILITY & GENERAL INFORMATION.

Acton's Modern Cookery for Private Families. Fcp. 8vo. 6s.
Black's Practical Treatise on Brewing. 8vo. 10s. 6d.
Buckton's Food and Home Cookery. Crown 8vo. 2s.
Bull on the Maternal Management of Children. Fcp. 8vo. 2s. 6d.
Bull's Hints to Mothers on the Management of their Health during the Period of Pregnancy and in the Lying-in Room. Fcp. 8vo. 2s. 6d.
Campbell-Walker's Correct Card, or How to Play at Whist. 32mo. 2s. 6d.
Crump's English Manual of Banking. 8vo. 15s.
Cunningham's Conditions of Social Well-Being. 8vo. 10s. 6d.
Handbook of Gold and Silver, by an Indian Official. 8vo. 12s. 6d.
Johnson's (W. & J. H.) Patentee's Manual. Fourth Edition. 8vo. 10s. 6d.
Longman's Chess Openings. Fcp. 8vo. 2s. 6d.

London, LONGMANS & CO.

Macleod's Economics for Beginners. Small crown 8vo. 2s. 6d.
— . Theory and Practice of Banking. 2 vols. 8vo. 26s.
— . Elements of Banking. Fourth Edition. Crown 8vo. 5s.
M'Culloch's Dictionary of Commerce and Commercial Navigation. 8vo. 63s.
Maunder's Biographical Treasury. Fcp. 8vo. 6s.
— Historical Treasury. Fcp. 8vo. 6s.
— Scientific and Literary Treasury. Fcp. 8vo. 6s.
— Treasury of Bible Knowledge. Edited by the Rev. J. Ayre, M.A. Fcp. 8vo. 6s.
— Treasury of Botany. Edited by J. Lindley, F.R.S. and T. Moore, F.L.S. Two Parts, fcp. 8vo. 12s.
— Treasury of Geography. Fcp. 8vo. 6s.
— Treasury of Knowledge and Library of Reference. Fcp. 8vo. 6s.
— Treasury of Natural History. Fcp. 8vo. 6s.
Pereira's Materia Medica, by Bentley and Redwood. 8vo. 25s.
Pewtner's Comprehensive Specifier; Building-Artificers' Work. Conditions and Agreements. Crown 8vo. 6s.
Pierce's Three Hundred Chess Problems and Studies. Fcp. 8vo. 7s. 6d.
Pole's Theory of the Modern Scientific Game of Whist. Fcp. 8vo. 2s. 6d.
Scott's Farm Valuer. Crown 8vo. 5s.
Smith's Handbook for Midwives. Crown 8vo. 5s.
The Cabinet Lawyer, a Popular Digest of the Laws of England. Fcp. 8vo. 9s.
West on the Diseases of Infancy and Childhood. 8vo. 18s.
Willich's Popular Tables for ascertaining the Value of Property. Post 8vo. 10s.
Wilson on Banking Reform. 8vo. 7s. 6d.
— on the Resources of Modern Countries 2 vols. 8vo. 24s.

MUSICAL WORKS BY JOHN HULLAH, LL.D.

Chromatic Scale, with the Inflected Syllables, on Large Sheet. 1s. 6d.
Card of Chromatic Scale. 1d.
Exercises for the Cultivation of the Voice. For Soprano or Tenor, 2s. 6d.
Grammar of Musical Harmony. Royal 8vo. 2 Parts, each 1s. 6d.
Exercises to Grammar of Musical Harmony. 1s.
Grammar of Counterpoint. Part I. super-royal 8vo. 2s. 6d.
Hullah's Manual of Singing. Parts I. & II. 2s. 6d.; or together, 5s.
Exercises and Figures contained in Parts I. and II. of the Manual. Books I. & II. each 8d.
Large Sheets, containing the Figures in Part I. of the Manual. Nos. 1 to 8 in a Parcel. 6s.
Large Sheets, containing the Exercises in Part I. of the Manual. Nos. 9 to 40, in Four Parcels of Eight Nos. each, per Parcel. 6s.
Large Sheets, the Figures in Part II. Nos. 41 to 52 in a Parcel, 9s.
Hymns for the Young, set to Music. Royal 8vo. 8d.
Infant School Songs. 6d.
Notation, the Musical Alphabet. Crown 8vo. 6d.
Old English Songs for Schools, Harmonised. 6d.
Rudiments of Musical Grammar. Royal 8vo. 3s.
School Songs for 2 and 3 Voices. 2 Books, 8vo. each 6d.
Time and Tune in the Elementary School. Crown 8vo. 2s. 6d.
Exercises and Figures in the same. Crown 8vo. 1s. or 2 Parts, 6d each.

London, LONGMANS & CO.

Spottiswoode & Co., Printers, New-street Square, London.

www.ingramcontent.com/pod-product-compliance
Lightning Source LLC
Chambersburg PA
CBHW020237170426
43202CB00008B/118